Heat Transfer

Peter von Böckh • Thomas Wetzel

Heat Transfer

Basics and Practice

 Springer

Prof. Dr. Peter von Böckh
Hedwig-Kettler-Str. 7
76137 Karlsruhe
Germany
bupvb@web.de

Prof. Dr.-Ing. Thomas Wetzel
Karlsruhe Institute of Technology
KIT
Kaiserstr. 12
76128 Karlsruhe
Germany
thomas.wetzel@kit.edu

ISBN 978-3-642-19182-4   e-ISBN 978-3-642-19183-1
DOI 10.1007/978-3-642-19183-1
Springer Heidelberg Dordrecht London New York

Library of Congress Control Number: 2011940209

Printed on acid-free paper

Springer is part of Springer Science+Business Media (www.springer.com)

Preface

This book is the English version of the fourth edition of the German book "Wärmeübertragung". I originally wrote the book based on my lecture notes. In my work with Asea Brown Boveri until 1991 I was closely involved with the design and development of heat exchangers for steam power plants. There the latest research results were required in the area of heat transfer, to develop new and more exact calculation procedures. In this business an accuracy of 0.5 % was required in order to be competitive.

However, although our young engineers were full theoretical knowledge about boundary layers, analogy theorems and a large number of calculation procedures, but could not design a very simple heat exchanger.

Later in my professorship at the University of Applied Sciences in Basel (Switzerland), I noticed that the most books for students on heat transfer were not up to date. Especially the American books with excellent didactic features, did not represent the state of the art in many fields. My lecture notes – and so this book – were then developed with the aim providing the students with state of the art correlations and enable them to really design and analyzing heat exchangers.

The VDI Heat Atlas presents the state of the art in heat transfer, but it is an expert's reference, too large and not instructive enough for students. It is used therefore frequently as a source in this book, but here we focus more on a didactic way of presenting the essentials of heat transfer along with many examples.

The first edition of this book was published in 2003. At the University of Applied Science in Basel after 34 lectures of 45 minutes the students could independently recalculate and design fairly complex heat exchangers, e.g. the cooling of a rocket combustion chambers, evaporators and condensers for heat pumps.

After my retirement Professor Thomas Wetzel, teaching Heat and Mass Transfer at Karlsruhe Institute of Technology (KIT), joined me as co-author. He is professor at the Institute of Thermal Process Engineering, the institute where large parts of the correlations in VDI Heat Atlas were developed. His professional background (heat transfer in molten semi-conductor materials, automotive compact heat exchangers and air conditioning, chemical process engineering) is complementary to my experience.

This book requires a basic knowledge of thermodynamics and fluid mechanics, e.g. first law of thermodynamics, hydraulic friction factors.

The examples in the book solved with *Mathcad* 14, can be down loaded from www.waermeuebertragung-online.de or www.springer.com/de/978-3-642-15958-9. The downloaded modules can be used for heat exchanger design. Also polinoms for material properties as described in Chapter 9 are programmed in *Mathcad* 14 and can so be implemented in other Mathcad 14 programs for the call of material properties of water, air and R134a.

We have to thank Prof. von Böckh's wife Brigitte for her help in completing this book. She spent a great deal of time on reviewing the book. She checked the correct size and style of letters, use of symbols, indices and composition. The appearance of the layout and legibility of the book is mainly her work.

Peter von Böckh with Thomas Wetzel, Karlsruhe, August 2011

Contents

6 Boiling heat transfer   171

7 Thermal radiation   189

8 Heat exchangers   215

List of Symbols

a	thermal difusivity	m²/s
$a = s_1/d$	dimensionless tube distance perpendicular to flow	-
A	flow cross-section, heat transfer area, surface area	m²
Bi	*Biot* number	-
B, b	width	m
$b = s_2/d$	dimensionless tube distance parallel to flow	-
C_{12}	radiation heat exchange coefficient	W/(m² K⁴)
C_s	*Stefan-Boltzmann*-constant of black bodies	5.67 W/(m² K⁴)
c	flow velocity	m/s
c_0	cross-flow inlet velocity	m/s
c_p	specific heat at constant pressure	J/(kg K)
D, d	diameter	m
d_A	bubble tear-off diameter	m
d_h	hydraulic diameter	m
F	force	N
F_s	gravity force	N
F_τ	sheer stress force	N
Fo	*Fourier* number	-
f_1, f_2	correction functions of heat transfer coefficients	-
f_A	correction function for tube arrangement in a tube bundle	-
$f_p f_n$	correction function for first row effect in tube bundles	-
g	gravitational acceleration	9,806 m/s²
Gr	*Grashof* number	-
H	height of a tube bundle	m
$H = m \cdot h$	enthalpy	J
h	*Planck*-constant	6,6260755 · 10⁻³⁴ J · s
h	specific enthalpy	J/kg, kJ/kg
h	fin height	m
i	number of tubes per tube row	-
$i_{\lambda,s}$	spectral specific intensity of black radiation	W/m³
k	overall heat transfer coefficient	W/(m² K)
k	*Boltzmann*-constant	1.380641 · 10⁻²³ J/K
$L' = A/U_{proj}$	flow length	m
$L' = \sqrt[3]{g/\nu^2}$	characteristic length of condensation	m
l	length	m
m	mass	kg

m	characteristic fin parameter	m^{-1}
$\dot{m}$	mass flow rate	kg/s
NTU	number of transfer units	-
Nu	*Nußelt* number	
n	number of tube rows, number of fins	-
p	pressure	Pa, bar
P	dimensionless temperature	-
Pr	*Prandtl* number	-
Q	heat	J
$\dot{Q}$	heat rate	W
$\dot{q}$	heat flux	W/m^2
R	individual gas constant	J/(kg K)
R_m	universal gas constant	J/(mol K)
R_a	mean roughness index	m
R_v	fouling resistance	(m^2 K)/W
r	radius	m
r	latent heat of evaporation	J/kg
R_1	ratio of heat capacity rate of fluid 1 to fluid 2	-
Ra	*Rayleigh* number	-
Re	*Reynolds* number	-
s_1	tube distance perpendicular to flow	m
s_2	tube distance parallel to flow	m
s	wall thickness	m
s_{Ri}	fin thickness	
T	absolute temperature	K
T_i	dimensionless temperature	-
t	time	s
t_{Ri}	fin distance	m
V	volume	m^3
$\dot{W} = \dot{m} \cdot c_p$	heat capacity rate	W/K
X	characteristic parameter for fin efficiency	-
x	steam quality	-
x, y, z	spacial coordinates	m
α_x	local heat transfer coefficients	W/(m^2 K)
α	mean heat transfer coefficients	W/(m^2 K)
α	absorptivity	-
β	thermal expansivity	1/T
β^0	bubble contact angle	°
δ	thickness of condensate film	m
δ_ϑ	thickness of thermal boundary layer	m
ε	emissivity	-
$\Delta\vartheta$	temperature difference	K
$\Delta\vartheta_{gr}, \Delta\vartheta_{kl}$	larger and smaller temperature difference at inlet and outlet	K

$\Delta\vartheta_m$	log mean temperature difference	K
ϑ	Celsius temperature	°C
ϑ', ϑ''	inlet resp. outlet temperature	°C
Θ	dimensionless temperature	-
η_{Ri}	fin efficiency	-
η	dynamc viscosity	kg/(m s)
ν	kinematc viscosity	m²/s
λ	thermal conduction	W/(m K)
λ	wave length	m
ρ	density	kg/m³
σ	surface tension	N/m
σ	*Stefan-Boltzmann*-constant	$5.6696 \cdot 10^{-8}$ W/(m² K⁴)
τ	sheer stress	N/m²
Ψ	hollow volume ratio, porosity	-
ξ	resistance factor	-

Indexes

1, 2, ..	state, fluid
12, 23, ...	change of state from 1 to 2
A	state at start of transient thermal conduction at time $t = 0$
A	bouyancy
a	outlet, outside
e	inlet
f	fluid
$f1, f2$	fluid 1, fluid 2
g	gas
i	inside
l	liquid
lam	laminar
m	mean value
m	middle
n	normal component of a vector
O	surface
r	radial component of a vector
Ri	fin
s	black body
turb	turbulent
W	wall
x	local value at location x, steam quality
x, y, z	x-, y- und z-components of a vector

1   Introduction and definitions

Heat transfer is a fundamental part of thermal engineering. It is the science of the rules governing the transfer of heat between systems of different temperatures. In thermodynamics, the heat transferred from one system to its surroundings is assumed as a given process parameter. This assumption does not give any information on how the heat is transferred and which rules determine the quantity of the transferred heat.

Heat transfer describes the dependencies of the heat transfer rate from a corresponding temperature difference and other physical conditions.

The thermodynamics terms *"control volume"* and *"system"* are also common in heat transfer. A system can be a material, a body or a combination of several materials or bodies, which transfer to or receive heat from another system.
The first two questions are:

- What is heat transfer?
- Where is heat transfer applied?

> *Heat transfer is the transport of thermal energy, due to a spacial temperature difference.*

> *If a spacial temperature difference is present within a system or between systems in thermal contact to each other, heat transfer occurs.*

The application of the science of heat transfer can be easily demonstrated with the example of a radiator design.

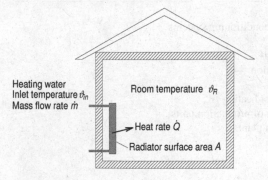

Heating water
Inlet temperature ϑ_{in}
Mass flow rate $\dot{m}$

Room temperature ϑ_R

Heat rate $\dot{Q}$

Radiator surface area A

Figure 1.1: Radiator design

To obtain a certain room temperature, radiators, in which warm water flows, are installed to provide this temperature. For the acquisition of the radiators, the architect defines the required heat flow rate, room temperature, heating water mass flow rate and temperature. Based on these data, the radiator suppliers make their offers. Is the designed radiator surface too small, temperature will be too low, the owner of the room will not be satisfied and the radiator must be replaced. Is the radiator surface too large, the room temperature will be too high. With throttling the heating water flow rate the required room temperature can be established. However, the radiator needs more material and will be too expensive, therefore it will not be ordered. The supplier with the correct radiator size will succeed. With experiments the correct radiator size could be obtained, but this would require a lot of time and costs. Therefore, calculation procedures are required, which allow the design of a radiator with an optimum size. For this example, the task of heat transfer analysis is to obtain the correct radiator size at minimum costs for the given parameters .

In practical design of apparatus or complete plants, in which heat is transferred, besides other technical sciences (thermodynamics, fluid mechanics, material science, mechanical design, etc.) the science of heat transfer is required. The goal is always to optimize and improve the products. The main goals are to:

* increase efficiency
* optimize the use of resources
* reach a minimum of environmental burden
* optimize product costs.

To reach these goals, an exact prediction of heat transfer processes is required.

> *To design a heat exchanger or a complete plant, in which heat is transferred, exact knowledge of the heat transfer processes is mandatory to ensure the greatest efficiency and the lowest total costs.*

Table 1.1 gives an overview of heat transfer applications.

Table 1.1: Area of heat transfer applications

Heating, ventilating and air conditioning systems
Thermal power plants
Refrigerators and heat pumps
Gas separation and liquefaction
Cooling of machines
Processes requiring cooling or heating
Heating up or cooling down of production parts
Rectification and distillation plants
Heat and cryogenic isolation
Solar-thermic systems
Combustion plants

1.1 Modes of heat transfer

Contrary to assured knowledge, most publications describe three *modes of heat transfer*: *thermal conduction, convection* and *thermal radiation*.
Nußelt, however, postulated in 1915, that only two modes of heat transfer exist [1.2] [1.3]. The publication of *Nußelt* states:

"In the literature it is often stated, heat emission of a body has three causes: radiation, thermal conduction and convection.

The separation of heat emission in thermal conduction and convection suggests that there would be two independent processes. Therefore, the conclusion would be: heat can be transferred by convection without the participation of thermal conduction. But this is not correct."

Heat transfer modes are thermal conduction and thermal radiation.

Figure 1.2 demonstrates the two modes of heat transfer.

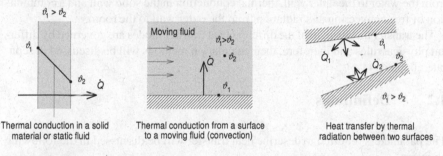

Thermal conduction in a solid material or static fluid

Thermal conduction from a surface to a moving fluid (convection)

Heat transfer by thermal radiation between two surfaces

Figure 1.2: Modes of heat transfer

1. *Thermal conduction* develops in materials when a spacial temperature gradient is present. With regard to calculation procedures there is a differentiation between static materials (solids and static fluids) and moving fluids. Heat transfer in static materials depends only on the spacial temperature gradient and material properties.

   Heat transfer between a solid wall and a moving fluid occurs by thermal conduction between the wall and the fluid and within the fluid. Furthermore, the transfer of enthalpy happens, which mixes areas of different temperatures. The heat transfer is determined by the thermal conductivity and the thickness of the boundary layer of the fluid, the latter is dependent on the flow and material parameters. In the boundary layer the heat is transferred by conduction.

   Because of the different calculation methods, the heat transfer between a solid wall and a fluid is called *convective heat transfer* or more concisely *convection*. A further differentiation is made between *free convection* and *forced convection*.

In free convection the fluid flow is generated by gravity due to the density difference caused by the spacial temperature gradient. At forced convection the flow is established by an external pressure difference.

2. *Thermal radiation* can occur without any intervening medium. All surfaces and gases consisting of more than two atoms per molecule of finite temperature, emit energy in the form of electromagnetic waves. Thermal radiation is a result of the exchange of electromagnetic waves between two surfaces of different temperature.

In the examples shown in Figure 1.3 the temperature ϑ_1 is larger than ϑ_2, therefore, the heat flux is in the direction of the temperature ϑ_2. In radiation both surfaces emit and absorb a heat flux, where the emission of the surface with the higher temperature ϑ_1 has a higher intensity.

Heat transfer may occur through combined thermal conduction and radiation. In many cases, one of the heat transfer modes is negligible. The heat transfer modes of the radiator, discussed at the beginning of this chapter are: forced convection inside from the water to the inner wall, thermal conduction in the solid wall and a combination of free convection and radiation from the outer wall to the room.

The transfer mechanism of the different heat transfer modes are governed by different physical rules and therefore, their calculation methods will be discussed in separate chapters.

1.2   Definitions

The parameters required to describe heat transfer will be discussed in the following chapters.

In this book the symbol ϑ is used for the temperature in Celsius and T for the absolute temperature.

1.2.1   Heat (transfer) rate and heat flux

The *heat rate*, also called *heat transfer rate* $\dot{Q}$ is the amount of heat transferred per unit time. It has the unit Watt **W**.

A further important parameter is the *heat flux density* $\dot{q} = \dot{Q}/A$, which defines the heat rate per unit area. Its unit is Watt per square meter **W/m²**.

1.2.2   Heat transfer coefficients and overall heat transfer coefficients

The description of the parameters, required for the definition of the heat flux density will be discussed in the example of a heat exchanger as shown in Figure 1.3. The heat exchanger consist of a tube that is installed in the center of a larger diameter tube.

A fluid with the temperature $\vartheta_1{}'$ enters the inner tube and will be heated up to the temperature $\vartheta_1{}''$. In the annulus a warmer fluid will be cooled down from the temperature $\vartheta_2{}'$ to the temperature $\vartheta_2{}''$. Figure 1.3 shows the temperature profiles in the fluids and in the wall of the heat exchanger.

The governing parameters for the heat rate transferred between the two fluids will be discussed now. The quantity of the transferred heat rate $\dot{Q}$ can be defined by the *heat transfer coefficient* α, the heat transfer surface area A and the temperature difference $\Delta\vartheta$.

> *The heat transfer coefficient defines the heat rate $\dot{Q}$ transferred per unit transfer area A and per unit temperature difference $\Delta\vartheta$.*

The unit of the heat transfer coefficient is $\mathbf{W/(m^2\,K)}$.

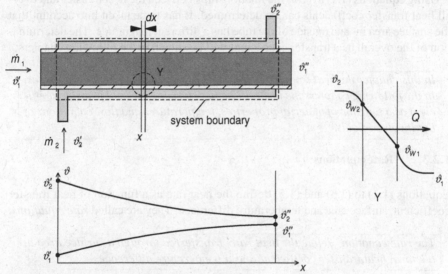

Figure 1.3: Temperature profile in the heat exchanger

With this definition, the finite heat rate through a finite surface element is:

$$\delta\dot{Q}_2 = \alpha_2 \cdot (\vartheta_2 - \vartheta_{W2}) \cdot dA_2 \tag{1.1}$$

$$\delta\dot{Q}_1 = \alpha_1 \cdot (\vartheta_{W1} - \vartheta_1) \cdot dA_1 \tag{1.2}$$

$$\delta\dot{Q}_W = \alpha_W \cdot (\vartheta_{W2} - \vartheta_{W1}) \cdot dA_W \tag{1.3}$$

The symbol $\delta\dot{Q}$ shows that the heat rate has an inexact differential, because the value of its integral depends on the heat transfer processes and path.

> *The integral of $\delta\dot{Q}$ is $\dot{Q}_{12}$ and not $\dot{Q}_2 - \dot{Q}_1$.*

Here the temperature differences were selected such that the heat rate has positive values. For a heat exchanger with a complete thermal insulation to the environment, the heat rate coming from fluid 2 must have the same value as the one transferred to fluid 1 and also have the same value as the heat rate through the pipe wall.

$$\delta \dot{Q}_1 = \delta \dot{Q}_2 = \delta \dot{Q}_W = \delta \dot{Q} \tag{1.4}$$

In most cases, the wall temperatures are unknown and the engineer is interested in knowing the total heat rate transferred from fluid 2 to fluid 1. For its determination the *overall heat transfer coefficient k* is required. It has the same unit as the heat transfer coefficient.

$$\delta \dot{Q} = k \cdot (\vartheta_2 - \vartheta_1) \cdot dA \tag{1.5}$$

Using equations (1.1) to (1.5) the relationships between the heat transfer and overall heat transfer coefficients can be determined. It has to be taken into account that the surface area in- and outside of the tube has a different magnitude. The determination of the overall heat transfer coefficient will be shown in the following chapters.

In this chapter the heat transfer coefficients are assumed to be known values. In the following chapters the task will be to determine the heat transfer coefficient as a function of material properties, temperatures and flow conditions of the involved fluids.

1.2.3   Rate equations

Equations (1.1) to (1.3) and (1.5) define the heat rate as a function of heat transfer coefficient, surface area and temperature difference. They are called *rate equations*.

The rate equations define the heat rate, transferred through a surface area at a known heat transfer coefficient and a temperature difference.

1.2.4   Energy balance equations

In heat transfer processes the first law of thermodynamics is valid without any restrictions. In most practical cases of heat transfer analysis, the mechanical work, friction, kinetic and potential energy are small compared to the heat rate. Therefore, for problems dealt with in this book, they are neglected. The *energy balance equation* of thermodynamics then simplifies to [1.1]:

$$\frac{dE_{CV}}{dt} = \dot{Q}_{CV} + \sum_e \dot{m}_e \cdot h_e - \sum_a \dot{m}_a \cdot h_a \tag{1.6}$$

The temporal change of energy in the control volume is equal to the total heat rate to the control volume and the enthalpy flows to and from the control volume. In most cases of heat transfer problems only one mass flow enters and leaves a control volume. The change of the enthalpy and energy in the control volume can be given as a function of the temperature. The heat rate is either transferred over the system boundary or originates from an internal source within the system boundary (e.g. electric heater, friction, chemical reaction). Equation (1.7) is presented here as it is mostly used for heat transfer problems:

$$V_{CV} \cdot \rho \cdot c_p \frac{d\vartheta}{dt} = \dot{Q}_{12} + \dot{Q}_{in} + \dot{m} \cdot (h_2 - h_1) \qquad (1.7)$$

In Equation (1.7) $\dot{Q}_{12}$ is the heat rate transferred over the system boundary and $\dot{Q}_{in}$ the heat rate originating from an internal source. For stationary processes the left side of Equation (1.7) has the value of zero:

$$\dot{Q}_{12} + \dot{Q}_{in} = \dot{m} \cdot (h_1 - h_2) = \dot{m} \cdot c_p \cdot (\vartheta_1 - \vartheta_2) \qquad (1.8)$$

The Equations (1.7) and (1.8) are called energy balance equations.

1.2.5 Log mean temperature difference

With known heat transfer coefficients, the heat rate at every location of the heat exchanger, shown in Figure 1.3, can be determined. In engineering, however, not the local but the total transferred heat is of interest. To determine the overall heat transfer rate, the local heat flux density must be integrated over the total heat transfer area. The total transferred heat rate is:

$$\dot{Q} = \int_0^A k \cdot (\vartheta_2 - \vartheta_1) \cdot dA \qquad (1.9)$$

The variation of the temperature in the surface area element dA can be calculated using the energy balance equation (1.8).

$$\delta\dot{Q} = \dot{m}_1 \cdot dh_1 = \dot{m}_1 \cdot c_{p1} \cdot d\vartheta_1 \qquad (1.10)$$

$$\delta\dot{Q} = -\dot{m}_2 \cdot dh_2 = -\dot{m}_2 \cdot c_{p2} \cdot d\vartheta_2 \qquad (1.11)$$

The temperature difference $\vartheta_2 - \vartheta_1$ will be replaced by $\Delta\vartheta$. The change of the temperature difference can be calculated from the change of the fluid temperatures.

$$d\Delta\vartheta = d\vartheta_2 - d\vartheta_1 = -\delta\dot{Q} \cdot \left(\frac{1}{\dot{m}_1 \cdot c_{p1}} + \frac{1}{\dot{m}_2 \cdot c_{p2}} \right) \qquad (1.12)$$

Equation (1.12) set in Equation (1.5) results in:

$$\frac{d\Delta\vartheta}{\Delta\vartheta} = -k \cdot \left(\frac{1}{\dot{m}_1 \cdot c_{p1}} + \frac{1}{\dot{m}_2 \cdot c_{p2}}\right) \cdot dA \tag{1.13}$$

Assuming that the overall heat transfer coefficient, the surface area and the specific heat capacities are constant, Equation (1.13) can be integrated. This assumption will never be fulfilled exactly. However, in practice the use of mean values has proven to be an excellent approach. The integration gives us:

$$\ln\left(\frac{\vartheta_2' - \vartheta_1'}{\vartheta_2'' - \vartheta_1''}\right) = k \cdot A \cdot \left(\frac{1}{\dot{m}_1 \cdot c_{p1}} + \frac{1}{\dot{m}_2 \cdot c_{p2}}\right) \tag{1.14}$$

With the assumptions above, Equations (1.10) and (1.11) can also be integrated.

$$\dot{Q} = \dot{m}_1 \cdot c_{p1} \cdot (\vartheta_1'' - \vartheta_1') \tag{1.15}$$

$$\dot{Q} = \dot{m}_2 \cdot c_{p2} \cdot (\vartheta_2' - \vartheta_2'') \tag{1.16}$$

In Equation (1.14) the mass flow rates and specific heat capacities can be replaced by the heat rate and fluid temperatures. This operation delivers:

$$\dot{Q} = k \cdot A \cdot \frac{\vartheta_2' - \vartheta_1' - \vartheta_2'' + \vartheta_1''}{\ln\dfrac{\vartheta_2' - \vartheta_1'}{\vartheta_2'' - \vartheta_1''}} = k \cdot A \cdot \Delta\vartheta_m \tag{1.17}$$

> *The temperature difference $\Delta\vartheta_m$ is the temperature difference relevant for the estimation of the heat rate. It is called the log mean temperature difference and is the integrated mean temperature difference of a heat exchanger.*

The *log mean temperature difference* is valid for the special case of the heat exchanger shown in Figure 1.3. For heat exchanger with parallel-flow, counterflow and if the temperature of one of the fluids remains constant (condensation and boiling) a generally valid log mean temperature difference can be given. For its formulation the temperature differences at the inlet and outlet of the heat exchangers are required. The greater temperature difference is $\Delta\vartheta_{gr}$, the smaller one is $\Delta\vartheta_{sm}$.

$$\Delta\vartheta_m = \frac{\Delta\vartheta_{gr} - \Delta\vartheta_{sm}}{\ln(\Delta\vartheta_{gr} / \Delta\vartheta_{sm})} \quad \text{if} \quad \Delta\vartheta_{gr} - \Delta\vartheta_{sm} \neq 0 \tag{1.18}$$

If the temperature differences at inlet and outlet are approximately identical, Equation (1.18) results in an indefinite value. For this case the log mean temperature difference is the average value of the inlet and outlet temperature differences.

$$\Delta\vartheta_m = (\Delta\vartheta_{gr} + \Delta\vartheta_{kl}) / 2 \quad \text{if} \quad \Delta\vartheta_{gr} - \Delta\vartheta_{kl} = 0 \tag{1.19}$$

The log mean temperature of a heat exchanger in which the flow of the fluids is perpendicular (cross-flow) or has changing directions will be discussed in Chapter 8.

1.2.6 Thermal conductivity

Thermal conductivity λ is a material property, which defines the magnitude of the heat rate that can be transferred per unit length in the direction of the flux and per unit temperature difference. Its unit is **W/(m K)**. The thermal conductivity of a material is temperature and pressure dependent.

Good electric conductors are usually also good thermal conductors, however, exceptions exist. Metals have a rather high thermal conductivity, liquids a smaller one and gases are "bad" heat conductors. In Figure 1.4 thermal conductivity of several materials is plotted versus temperature.

The thermal conductivity of most materials does not vary much at a medium temperature change. Therefore, they are suitable for calculation with constant mean values.

1.3 Methodology of solving problems

This chapter originates from [1.5], with small changes. For *solving problem*s of heat transfer usually, directly or indirectly, the following basic laws and principles are required:

- law of *Fourier*
- laws of heat transfer
- conservation of mass principle
- conservation of energy principle (first law of thermodynamics)
- second law of thermodynamics
- *Newton*'s second law of motion
- momentum equation
- similarity principles
- friction principles

Besides profound knowledge of the basic laws, the engineer has to know the *methodology*, i.e. *how* to apply the above mentioned basic laws and principles to concrete problems. It is of great importance to learn a systematic analysis of problems. This consist mainly of six steps as listed below. They are proven in practice and can, therefore, highly be recommended.

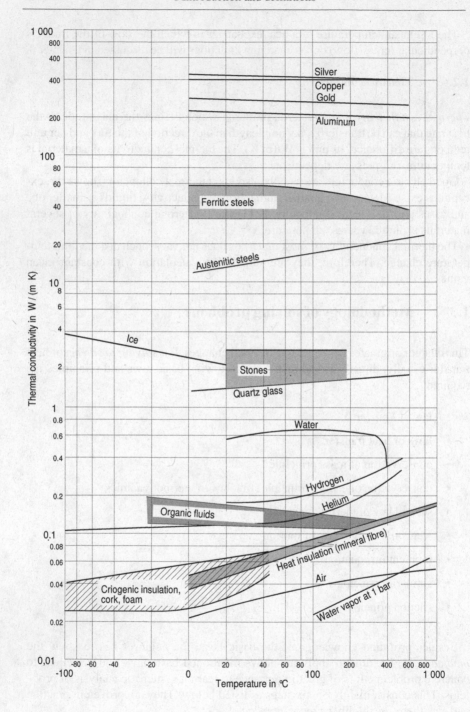

Fig. 1.4: Thermal conductivity of materials versus temperature [1.5]

1. What is given?

Analyze, what is known about the problem you have to investigate. Note all parameters which are given or may be necessary for further considerations.

2. What are you looking for?

At the same time as the first step, analyze which parameter has to be determined and which questions have to be answered.

3. How is the system defined?

Make schematic sketches of the system and decide which types of system boundaries are best for the analysis.

- Define the system boundary(ies) clearly!

Identify the *transactions* between systems and environment.
State which changes of state or processes are acting on or are passing through the system.

- Create clear system schematics!

4. Assumptions

Consider how the system can be modelled as simple as possible, make *simplifying assumptions*. Specify the boundary values and assumptions.
Check if *idealizing assumptions* are possible, e.g. physical properties determined with mean temperatures and negligible heat losses assumed as perfect heat insulation.

5. Analysis

Collect all necessary material properties. Some of them may be given in the appendices of this book. If not, a research in the literature is requested (e.g. VDI Heat Atlas [1.7]).
Take into account the idealizing and simplifying assumptions, formulate the heat balance and rate equations.

Recommendation: Finish all formulations, transformations, simplifications and solutions in symbolic equations, before inserting numeric values.
Check the equations and data of correctness of units and dimensions before the numeric evaluation is started.
Check the order of magnitude of the results and the correctness of algebraic signs.

6. Discussion

Discuss the results and its basic aspects, comment on the main results and their correlations.

Do remember: Step 3 is of greatest importance, as it clarifies how the analysis was performed and step 4 is as important, as it determines the quality and validity range of the results. The examples in this book are treated according to the above described methodology. The definition of the examples is according to steps 1 and 2, the solution starts with step 3.

EXAMPLE 1.1: Determination of the heat rate, temperature and heat transfer surface area

The counterflow heat exchanger is a pipe, installed concentrically in a pipe of lager diameter. The fluid in the pipe and in the annulus is water in this example. The mass flow rate in the pipe is 1 kg/s and the inlet temperature 10 °C . In the annulus we have a mass flow rate of 2 kg/s and the water is cooled from 90 °C down to 60 °C The overall heat transfer coefficient is 4000 W/(m² K). The specific heat capacity of the water in the pipe is 4.182 kJ/(kg K) and that in the annulus 4.192 kJ/(kg K).

Find

The heat rate, the outlet temperature of the water in the pipe and the required heat transfer surface area.

Solution

Schematic   See sketch.

Assumptions

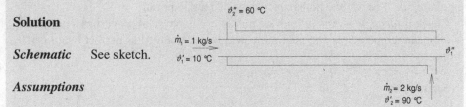

- The heat exchanger does not transfer any heat to the environment.
- The process is stationary.

Analysis

The heat rate transferred from the water in the annulus to the water in the tube can be calculated with the energy balance Equation (1.15).

$$\dot{Q} = \dot{m}_2 \cdot c_{p2} \cdot (\vartheta_2' - \vartheta_2'') =$$
$$= 2 \cdot \text{kg/s} \cdot 4192 \ \cdot \text{J/(kg} \cdot \text{K)} \cdot (90 - 60) \cdot \text{K} = \textbf{251.52 kW}$$

The outlet temperature of the water determined with Equation (1.16) is:

$$\vartheta_1'' = \vartheta_1' + \frac{\dot{Q}}{\dot{m}_1 \cdot c_{p1}} = 10\ °C + \frac{251520\ W}{1 \cdot kg/s \cdot 4182 \cdot J/(kg \cdot K)} = \mathbf{70.1\ °C}$$

For the determination of the required surface area Equations (1.17) and (1.18) are used. First with Equation (1.18) the log mean temperature difference $\Delta\vartheta_m$ is calculated. At the inlet of the tube the large temperature difference has the value of 50 K, the small one at the outlet 19.9 K.

$$\Delta\vartheta_m = \frac{\Delta\vartheta_{gr} - \Delta\vartheta_{kl}}{\ln(\Delta\vartheta_{gr} / \Delta\vartheta_{kl})} = \frac{(50 - 19.9) \cdot K}{\ln(50/19.9)} = 32.6\ K$$

With Equation (1.17) the required surface area results as:

$$A = \frac{\dot{Q}}{k \cdot \Delta\vartheta_m} = \frac{251520\ W}{4000\ \cdot W/(m^2 \cdot K) \cdot 32.6 \cdot K} = \mathbf{1.93\ m^2}$$

Discussion

With a known heat rate and energy balance equations the temperature change of the liquids can be calculated. To determine the required heat exchanger surface area the rate equation is required, however first the overall heat transfer coefficient has to be known. Its calculation will be discussed in the following chapters. This example shows that with water, a relatively small surface area is sufficient to transfer a fairly large heat rate.

EXAMPLE 1.2: Determination of the outlet temperature

In the heat exchanger in Example 1.1, the inlet temperature of the water in the tube has changed from 10 °C to 25 °C. The mass flow rates, material properties and the inlet temperature of the water into the annulus remain the same as in Example 1.1.

Find

The outlet temperatures and the heat rate.

Solution

Assumptions

• The heat transfer coefficient in the whole heat exchanger is constant.
• The process is stationary.

Analysis

The Equations (1.15) to (1.17) provide three independent equations for the determination of the three unknown values: $\dot{Q}$, ϑ_1'' and ϑ_2''. The energy balance equations for the two mass flow rates are:

$$\dot{Q} = \dot{m}_1 \cdot c_{p1} \cdot (\vartheta_1'' - \vartheta_1') \quad \text{and} \quad \dot{Q} = \dot{m}_2 \cdot c_{p2} \cdot (\vartheta_2' - \vartheta_2'')$$

The rate equation is:

$$\dot{Q} = k \cdot A \cdot \frac{\vartheta_2' - \vartheta_1' - (\vartheta_2' - \vartheta_1'')}{\ln \dfrac{\vartheta_2' - \vartheta_1'}{\vartheta_2' - \vartheta_1''}} = k \cdot A \cdot \Delta\vartheta_m$$

The temperatures in the numerator of the above equation can be replaced by the heat rate divided by the product of mass flow rate and heat capacity. The above three equations deliver:

$$\frac{\vartheta_2' - \vartheta_1'}{\vartheta_2' - \vartheta_1''} = e^{k \cdot A \cdot \left(\frac{1}{\dot{m}_1 \cdot c_{p1}} - \frac{1}{\dot{m}_2 \cdot c_{p2}} \right)} = e^{4000 \cdot 1.93 \cdot \frac{W}{K} \cdot \left(\frac{1}{1 \cdot 4182} - \frac{1}{2 \cdot 4192} \right) \frac{K}{W}} = 2.518$$

With the energy balance equation the temperature ϑ_2'' can be given as a function of the temperature ϑ_1'' and inserted in the equation above. The equation solved for ϑ_2'' delivers.

$$\vartheta_2'' = \vartheta_2' - \frac{\dot{m}_1 \cdot c_{p1}}{\dot{m}_2 \cdot c_{p2}} \cdot (\vartheta_1'' - \vartheta_1') = 90\ °C - 0.4988 \cdot (\vartheta_1'' - 25°C) =$$

$$= 102.47°C - 0.4988 \cdot \vartheta_1''$$

$$\vartheta_1'' = \frac{2.522 \cdot \vartheta_2' + \vartheta_1' - 102.47°C}{2.0232} = \frac{2.522 \cdot 90°C + 25\ °C - 102.47°C}{2.0232} = \mathbf{73.9°C}$$

The result for temperature ϑ_1'' is:

$$\vartheta_2'' = 102.47°C - 0.4988 \cdot \vartheta_1'' = \mathbf{65.6°C}$$

The heat rate can be determined with the energy balance equation.

$$\dot{Q} = \dot{m}_1 \cdot c_{p1} \cdot (\vartheta_1'' - \vartheta_1') = 1 \cdot \frac{kg}{s} \cdot 4182 \cdot \frac{J}{kg \cdot K} \cdot (73.9 - 25) \cdot K = \mathbf{204.5\,kW}$$

Discussion

With the energy balance and the rate equations, the outlet temperatures of an already designed heat exchanger (of which the physical dimensions are known) can be calculated. With increased inlet temperature of the cold water in the pipe, the outlet temperature increases but the heat rate decreases as the temperature changes of both water flows decrease.

2   Thermal conduction in static materials

Thermal conduction in static materials is a heat transfer process in solids or static fluids. The carriers of the energy transfer can be molecules, atoms, electrons and phonons. The latter are energy quantums of elastic waves, in nonmetallic and metallic solids. Electrons transfer heat in metals, both in solid and fluid state.

> *Heat transfer will occur in any static material, as soon as a spacial tempera-ture gradient exists.*

In this chapter, only heat transfer in solids and static fluids will be discussed. Thermal conduction in moving fluids, called convection, will be treated in Chapters 3 to 6. Thermal conduction in static fluids is a rare phenomenon in practice, as the density differences in the fluid, caused by the temperature differences, generate gravity-driven flows.

Heat transfer with a constant heat flux and a steady-state spacial temperature distribution is called *steady-state thermal conduction*. If a body is heated up or cooled down, there is a transient change of the heat rate and also of the spacial temperature distribution. This process is called *transient thermal conduction*.

2.1   Steady-state thermal conduction

The heat flux, caused by a temperature difference in a material, is defined by the law of *Fourier*.

$$\dot{q} = -\lambda \cdot \nabla \vartheta = -\lambda \cdot \frac{d\vartheta}{dr} \tag{2.1}$$

The spacial coordinate is r. The heat flux is proportional to the thermal conductivity and to the spacial temperature gradient. It is always contrary to the direction of the temperature gradient. According to Equation (2.1) the vector of the heat flux is per-pendicular to any isothermal surface. Alternatively, the law of *Fourier* can be given as:

$$\dot{q}_n = -\lambda \cdot \frac{d\vartheta}{dn} \tag{2.2}$$

Where $\dot{q}_n$ is the normal component of the heat flux vector through an arbitrary surface and n the normal component of the space vector at this surface.

The heat rate through the surface area is:

$$\dot{Q} = \int_A \dot{q}_n \cdot dA \qquad (2.3)$$

The thermal conductivity is a function of the temperature and the surface area can be a more or less complicated function of the spacial coordinates, therefore the solution of the integral can be very complicated or even impossible. For many technical applications the thermal conductivity can be taken as constant with a mean value. In bodies with simple geometrical shapes, Equation (2.3) can be solved.

2.1.1 Thermal conduction in a plane wall

Figure 2.1 shows a plane wall with the thickness s and the thermal conductivity λ. At the top and bottom the wall is thermally completely insulated. Heat transfer is possible only in x direction, therefore the problem is one dimensional. The surface area A of the wall, through which the heat flux passes, is constant. Equation 2.1 can be given as:

$$\dot{Q} = -\lambda \cdot A \cdot \frac{d\vartheta}{dx} \qquad (2.4)$$

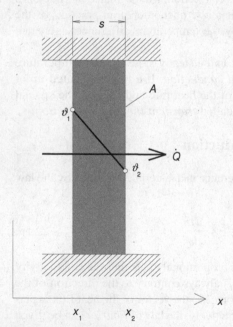

Figure 2.1: Thermal conduction in a plane wall

Due to the thermal insulation of the top and bottom of the wall sideways leakage of heat is impossible. Assuming a constant thermal conductivity, there exists only a heat rate in x direction and Equation (2.4) can be integrated.

$$\int\limits_{x_1}^{x_2}\dot{Q}\cdot dx = \int\limits_{\theta_1}^{\theta_2}-\lambda\cdot A\cdot d\vartheta \qquad (2.5)$$

$$\dot{Q} = \frac{\lambda}{s}\cdot A\cdot(\vartheta_1-\vartheta_2) \qquad (2.6)$$

In a plane wall with constant thermal conductivity, the temperature distribution is linear. With the definition (1.1) of the heat transfer coefficient we receive:

$$\alpha = \lambda / s \qquad (2.7)$$

> *The heat transfer coefficient of a plane wall with constant thermal conductivity is the thermal conductivity divided by the wall thickness.*

To establish constant temperatures on both sides of the wall, as shown in Figure 2.1, on the one side a heat source emitting a constant heat flux and a heat sink on the other side absorbing this heat flux, are required. This could be, for example, on one side a moving warm fluid, that delivers the heat rate, and a cold fluid on the side, receiving the heat rate. This is the case in heat exchangers, where heat is transferred through a solid wall from a fluid 1 to a fluid 2.

Figure 2.2 shows the plane wall of a heat exchanger, in which the heat is transferred from a warm fluid with the temperature ϑ_{f1} and a given heat transfer coefficient α_{f1} to a cold fluid with the temperature ϑ_{f2} and a given heat transfer coefficient α_{f2}. Here the fluid heat transfer coefficients are assumed as known values. Their determination will be discussed in Chapters 3 to 7.

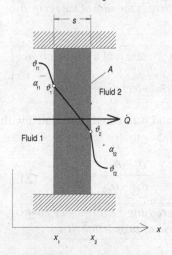

Figure 2.2: Determination of the overall heat transfer coefficient

By definition, the heat transfer coefficient represents the heat rate, that can be transferred at a certain temperature difference. The heat rates from fluid 1 to the wall, through the wall and from the wall to fluid 2 can be calculated with the rate equations.

$$\dot{Q} = A \cdot \alpha_{f1} \cdot (\vartheta_{f1} - \vartheta_1)$$
$$\dot{Q} = A \cdot \alpha_w \cdot (\vartheta_1 - \vartheta_2)$$
$$\dot{Q} = A \cdot \alpha_{f2} \cdot (\vartheta_2 - \vartheta_{f2})$$

(2.8)

With the overall heat transfer coefficient k the heat flux from fluid 1 to fluid 2 can be directly determined.

$$\dot{Q} = A \cdot k \cdot (\vartheta_{f1} - \vartheta_{f2})$$

(2.9)

To calculate the overall heat transfer coefficient, first the wall temperatures ϑ_1 and ϑ_2 must be determined with Equation (2.8).

$$\vartheta_1 = \vartheta_{f1} - \frac{\dot{Q}}{A \cdot \alpha_{f1}} \qquad\qquad \vartheta_2 = \vartheta_{f2} + \frac{\dot{Q}}{A \cdot \alpha_{f2}}$$

(2.10)

Inserting Equations 2.10 in 2.8 delivers:

$$\dot{Q} \cdot \left(\frac{1}{\alpha_{f1}} + \frac{1}{\alpha_w} + \frac{1}{\alpha_{f2}} \right) = A \cdot (\vartheta_{f1} - \vartheta_{f2})$$

(2.11)

The overall heat transfer coefficient is:

$$\frac{1}{k} = \frac{1}{\alpha_{f1}} + \frac{1}{\alpha_w} + \frac{1}{\alpha_{f2}}$$

(2.12)

The reciprocal of the overall heat transfer coefficient is the sum of the reciprocals of the heat transfer coefficients.

The reciprocal of the product of heat transfer coefficient and transfer surface area $1/(\alpha \cdot A)$ is a thermal resistance.
The thermal resistances have to be added as serial electric resistances.

The temperature differences between the fluid and wall can be calculated with Equations (2.8) and (2.9).

$$\frac{\vartheta_{f1} - \vartheta_1}{\vartheta_{f1} - \vartheta_{f2}} = \frac{k}{\alpha_{f1}} \qquad \frac{\vartheta_1 - \vartheta_2}{\vartheta_{f1} - \vartheta_{f2}} = \frac{k}{\alpha_w} \qquad \frac{\vartheta_2 - \vartheta_{f2}}{\vartheta_{f1} - \vartheta_{f2}} = \frac{k}{\alpha_{f2}}$$

(2.13)

For the transfer of a certain heat rate, with decreasing heat transfer coefficient, the required temperature difference increases proportionally.

EXAMPLE 2.1: Determination of the wall heat transfer coefficient, overall heat transfer coefficient and the wall temperatures

Inside a room the air temperature is 22 °C, that outside 0 °C. The wall has a thickness of 400 mm and a thermal conductivity of 1 W/(m K). The heat transfer coefficient of the air on both sides has the value of 5 W/(m² K).

Find

The heat transfer coefficient of the wall, the overall heat transfer coefficient, heat flux and the wall temperatures.

Solution

Schematic   See sketch.

Assumptions

- The thermal conductivity in the wall is constant.
- No heat losses through the sides of the wall.
- The process is stationary.

Analysis

With Equation (2.7) the heat transfer coefficient in the wall can be determined.

$$\alpha_W = \frac{\lambda}{s} = \frac{1 \cdot W/(m \cdot K)}{0.4 \cdot m} = 2.5 \; \frac{W}{m^2 \cdot K}$$

The overall heat transfer coefficient is being calculated with Equation (2.12).

$$k = \left(\frac{1}{\alpha_{f1}} + \frac{1}{\alpha_W} + \frac{1}{\alpha_{f2}} \right)^{-1} = \left(\frac{1}{5} + \frac{1}{2.5} + \frac{1}{5} \right)^{-1} \cdot \frac{W}{m^2 \cdot K} = 1.25 \; \frac{W}{m^2 \cdot K}$$

To determine the heat flux, Equation (2.9) can be used.

$$\dot{q} = \dot{Q}/A = k \cdot (\vartheta_{f1} - \vartheta_{f2}) = 1.25 \cdot W/(m^2 \cdot K) \cdot (22 - 0) \cdot K = 27.5 \; W/m^2$$

The wall temperatures can be calculated with Equation (2.13) or with Equation (2.8). The temperature of the wall inside is determined with Equation (2.13), that outside with Equation (2.8).

$$\vartheta_1 = \vartheta_{f1} - (\vartheta_{f1} - \vartheta_{f2}) \frac{k}{\alpha_{f1}} = 22 \; °C - (22 - 0) \cdot K \cdot \frac{1.25}{5} = 16.5 \, °C$$

$$\vartheta_2 = \dot{q}/\alpha_{f2} + \vartheta_{f2} = 27.5/5 \cdot K + 0 \; °C = 5.5 \, °C$$

Discussion

The calculation demonstrates that the smallest heat transfer coefficient has the largest influence on the overall heat transfer coefficient. The largest temperature difference is in the material with the lowest heat transfer coefficient, here in the wall with 11 K.

2.1.2   Heat transfer through multiple plane walls

Plane walls are often consisting of multiple layers (wall of a house, insulation of a refrigerator). Figure 2.3 shows a wall with n layers of different thicknesses and thermal conductivities.

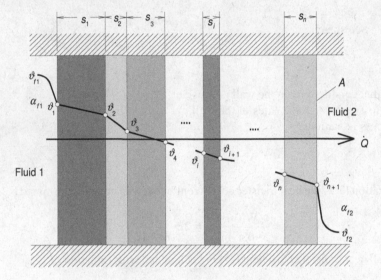

Figure 2.3: Heat transfer through a wall consisting of multiple plane layers

The heat transfer coefficient of each layer can be determined with Equation (2.7).

$$\alpha_i = \lambda_i / s_i \tag{2.14}$$

The overall heat transfer coefficient can be determined with the same calculation procedure as used for a plane wall.

$$\frac{1}{k} = \frac{1}{\alpha_{f1}} + \sum_{i=1}^{n} \frac{1}{\alpha_{Wi}} + \frac{1}{\alpha_{f2}} = \frac{1}{\alpha_{f1}} + \sum_{i=1}^{n} \frac{s_i}{\lambda_i} + \frac{1}{\alpha_{f2}} \tag{2.15}$$

Equation (2.13) delivers for the temperature differences in the wall layers:

$$\frac{\vartheta_{f1} - \vartheta_1}{\vartheta_{f1} - \vartheta_{f2}} = \frac{k}{\alpha_{f1}} \qquad \frac{\vartheta_i - \vartheta_{i+1}}{\vartheta_{f1} - \vartheta_{f2}} = \frac{k}{\alpha_{Wi}} \qquad \frac{\vartheta_2 - \vartheta_{f2}}{\vartheta_{f1} - \vartheta_{f2}} = \frac{k}{\alpha_{f2}} \tag{2.16}$$

EXAMPLE 2.2: Determination of the insulation layer thickness of a house wall

The wall of a house consists of an outer brick layer of 240 mm thickness and an inner layer of 120 mm thickness. Between the two walls there is mineral fibre insulation layer. The thermal conductivity of the inner and outer wall is 1 W/(m K), that of the insulation 0.035 W/(m K). The overall heat transfer coefficient of the multiple layer house wall shall not exceed 0.3 W/(m² K).

Find

The required insulation thickness.

Solution

Schematic   See sketch.

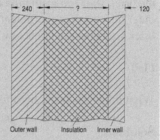

Outer wall   Insulation   Inner wall

Assumptions

- The thermal conductivities of all layers are areally homogeneous and independent from the temperature.
- No heat losses on sides of the wall.

Analysis

The overall heat transfer coefficient is given with Equation (2.15).

$$\frac{1}{k} = \sum_{i=1}^{n} \frac{s_i}{\lambda_i} = \frac{s_1}{\lambda_1} + \frac{s_2}{\lambda_2} + \frac{s_3}{\lambda_2}$$

In this example the overall heat transfer coefficient is known, the thickness of the insulation layer s_2 is to be determined. Therefore, with the above equation s_2 can be calculated.

$$s_2 = \left(\frac{1}{k} - \frac{s_1}{\lambda_1} - \frac{s_3}{\lambda_3} \right) \cdot \lambda_2 =$$

$$= \left(\frac{1}{0{,}3} - \frac{0.24}{1} - \frac{0.12}{1} \right) \cdot \frac{m^2 \cdot K}{W} \cdot 0.035 \cdot \frac{W}{m^2 \cdot K} = \mathbf{0.104\,m}$$

Discussion

The insulation layer is the main heat transfer resistance. Its heat transfer coefficient of 0.337 W/(m²·K) is only 12 % higher as the overall heat transfer coefficient.

EXAMPLE 2.3: Determination of the insulation layer thickness and the wall temperature of a cold store

The wall of a cold store consists of an outer brick wall of 200 mm thickness, an insulation layer and an inner plastic covering of 5 mm thickness. The thermal conductivities: brick wall 1 W/(m K), plastic 1,5 W/(m K), insulation 0.04 W/(m K). The cold storage temperature is −22 °C. The heat transfer coefficient of air in the cold storage is 8 W/(m² K). At an outside temperature of 35 °C it must be made sure, that between the brick wall and insulation layer no dew formation occurs. This requires that at an outer heat transfer coefficient of 5 W/(m² K) the temperature on the inner side of the brick wall does not fall below 32 °C.

Find

What thickness is required for the insulation layer?

Solution

Schematic　　See sketch.

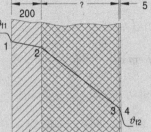

Brick wall　Insulation　Plastic cover

Assumptions

- The thermal conductivities of all layers are areally homogeneous and independent from the temperature.
- No heat leakage at side of the wall.

Analysis

After conversion of Equation (2.16) the overall heat transfer coefficient, at which no dew formation occurs, is:

$$k = \frac{\vartheta_{f_1} - \vartheta_2}{(\vartheta_{f1} - \vartheta_{f2}) \cdot \left(\dfrac{1}{\alpha_{f1}} + \dfrac{s_1}{\lambda_1} \right)} = \frac{(35-32) \cdot \text{K}}{(35+22) \cdot \text{K} \cdot \left(\dfrac{1}{5} + \dfrac{0.2}{1} \right) \cdot \dfrac{\text{m}^2 \cdot \text{K}}{\text{W}}} = 0.132 \frac{\text{W}}{\text{m}^2 \cdot \text{K}}$$

The thickness of the insulation layer can be calculated as in example 2.1.

$$s_2 = \left(\frac{1}{k} - \frac{1}{\alpha_{f1}} - \frac{s_1}{\lambda_1} - \frac{s_3}{\lambda_3} - \frac{1}{\alpha_{f2}} \right) \cdot \lambda_2 = \textbf{0.283 m}$$

Discussion

The main heat transfer resistance and thus the largest temperature difference is in the insulation layer. With the insulation layer thickness the overall heat transfer coefficient and the wall temperatures can be influenced.

2.1.3   Thermal conduction in a hollow cylinder

In a plane wall the surface area for the heat rate is constant. In a hollow cylinder (tube wall) the surface area changes with the radius r, thus A is a function of r. Figure 2.4 demonstrates the thermal conduction in a hollow cylinder.

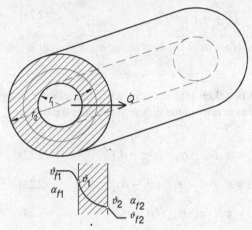

Figure 2.4: Thermal conduction in a hollow cylinder

The heat rate through the cylinder wall is constant. As the cylinder surface area changes with the radius, the heat flux also changes. The surface area as a function of the radius inserted in Equation (2.3) delivers:

$$\dot{Q} = -\lambda \cdot A(r) \cdot \frac{d\vartheta}{dr} = -\lambda \cdot \pi \cdot 2 \cdot r \cdot l \cdot \frac{d\vartheta}{dr} \tag{2.17}$$

The separation of variables results in:

$$\frac{dr}{r} = -\lambda \cdot \frac{2 \cdot \pi \cdot l}{\dot{Q}} \cdot d\vartheta \tag{2.18}$$

With the assumption, that the thermal conductivity in the wall is constant and a cylindrically symmetric heat flux occurs, Equation (2.18) can be integrated. The result for the heat rate is:

$$\dot{Q} = \lambda \cdot \frac{2 \cdot \pi \cdot l}{\ln(r_a / r_i)} \cdot (\vartheta_1 - \vartheta_2) \tag{2.19}$$

To determine the heat transfer coefficient, a reference surface area must first be defined. In Europe it is common to use the outer wall surface area as reference, but if necessary or more convenient for the calculation the inner wall surface area can also be selected. Equation (2.19) will be converted such that the reference surface areas is that of the outer wall surface. Furthermore, the radius will be replaced by the more commonly used diameter.

$$\dot{Q} = \frac{2 \cdot \lambda}{d_a \cdot \ln(d_a / d_i)} \cdot \pi \cdot l \cdot d_a \cdot (\vartheta_1 - \vartheta_2) = \frac{2 \cdot \lambda}{d_a \cdot \ln(d_a / d_i)} \cdot A_a \cdot (\vartheta_1 - \vartheta_2) \quad (2.20)$$

The heat transfer coefficient related to the outer wall surface area then reads:

$$\alpha_{Wa} = \frac{2 \cdot \lambda}{d_a \cdot \ln(d_a / d_i)} \quad (2.21)$$

> *It is very important to define which surface area the heat transfer coefficient is related to.*

The different surface areas have to be taken into account when calculating the overall heat transfer coefficient. For the heat rates at the different surfaces the following rate equations exist:

$$\dot{Q} = A_1 \cdot \alpha_{f1} \cdot (\vartheta_{f1} - \vartheta_1) = \pi \cdot l \cdot d_1 \cdot \alpha_{f1} \cdot (\vartheta_{f1} - \vartheta_1) \quad (2.22)$$

$$\dot{Q} = A_2 \cdot \alpha_{f2} \cdot (\vartheta_2 - \vartheta_{f2}) = \pi \cdot l \cdot d_2 \cdot \alpha_{f2} \cdot (\vartheta_2 - \vartheta_{f2}) \quad (2.23)$$

$$\dot{Q} = A_2 \cdot \alpha_{Wa} \cdot (\vartheta_1 - \vartheta_2) = \pi \cdot l \cdot d_2 \cdot \alpha_{Wa} \cdot (\vartheta_1 - \vartheta_2) \quad (2.24)$$

$$\dot{Q} = A_2 \cdot k \cdot (\vartheta_{f1} - \vartheta_{f2}) = \pi \cdot l \cdot d_2 \cdot k \cdot (\vartheta_{f1} - \vartheta_{f2}) \quad (2.25)$$

Note, that according to the first law of thermodynamics, the heat rate crossing the different surface areas must be the same, as long as a steady state is assumed for the whole cylinder volume. After conversions, the heat rate results in:

$$\dot{Q} = A_2 \cdot (\vartheta_{f1} - \vartheta_{f2}) \cdot \left[\frac{d_2}{d_1} \cdot \frac{1}{\alpha_{f1}} + \frac{1}{\alpha_{Wa}} + \frac{1}{\alpha_{f2}} \right]^{-1} \quad (2.26)$$

With this, the overall heat transfer coefficient related to the outside surface area of the cylinder, i.e. to the outer diameter d_2, results as:

$$\frac{1}{k} = \frac{d_2}{d_1} \cdot \frac{1}{\alpha_{f1}} + \frac{1}{\alpha_{Wa}} + \frac{1}{\alpha_{f2}} = \frac{d_2}{d_1} \cdot \frac{1}{\alpha_{f1}} + \frac{d_2}{2 \cdot \lambda} \cdot \ln\left(\frac{d_2}{d_1}\right) + \frac{1}{\alpha_{f2}} \quad (2.27)$$

By relating the overall heat transfer coefficient to the inner wall surface area, i.e. to the inner diameter d_1, the overall heat transfer will be:

$$\frac{1}{k_1} = \frac{1}{\alpha_{f1}} + \frac{d_1}{d_2} \cdot \frac{1}{\alpha_{Wa}} + \frac{d_1}{d_2} \cdot \frac{1}{\alpha_{f2}} = \frac{1}{\alpha_{f1}} + \frac{d_1}{2 \cdot \lambda} \cdot \ln\left(\frac{d_2}{d_1}\right) + \frac{d_1}{d_2} \cdot \frac{1}{\alpha_{f2}} \quad (2.28)$$

Both equations deliver the same heat rate, because it results from the multiplication of the overall heat transfer coefficient with the reference surface area. By using the wrong reference surface, big differences can result in the determination of the required surface area of the heat exchanger.

> *It is important that the reference surface area of the heat transfer coefficient is defined.*

The temperature differences in the fluid and in the wall are:

$$\frac{\vartheta_{f1} - \vartheta_1}{\vartheta_{f1} - \vartheta_{f2}} = \frac{d_a}{d_i} \cdot \frac{k}{\alpha_{f1}} \qquad \frac{\vartheta_1 - \vartheta_2}{\vartheta_{f1} - \vartheta_{f2}} = \frac{k}{\alpha_{Wa}} \qquad \frac{\vartheta_2 - \vartheta_{f2}}{\vartheta_{f1} - \vartheta_{f2}} = \frac{k}{\alpha_{f2}} \qquad (2.29)$$

For very thin tube walls or for rough estimates the tube wall can be handled as a plane wall. For this approximation the tube wall heat transfer coefficient is:

$$\alpha_{Wa} = \frac{2 \cdot \lambda}{d_a \cdot \ln(d_a / d_i)} \approx \frac{\lambda}{s} = \frac{2 \cdot \lambda}{d_a - d_i} \qquad (2.30)$$

EXAMPLE 2.4: Overall heat transfer coefficient in a heat exchanger tube

In a high pressure reheater the water flow in the tube is heated by steam condensing on the outer wall. The heat transfer coefficient in the tube is $15\,000$ W/(m² K) and outside $13\,000$ W/(m² K). The tube has an outer diameter of 15 mm and a wall thickness of 2.3 mm. The thermal conductivity of the tube is 40 W/(m K).

Find

a) The overall heat transfer coefficient related to the outer tube surface area.
b) The overall heat transfer coefficient related inner tube surface area.
c) The error when calculating the wall heat transfer coefficient with Equation (2.30).

Solution

Schematics   See Figure 2.4.

Assumptions

• The heat conductivity in the wall is constant.
• The temperatures inside and outside of the wall are constant.

Analysis

a) The overall heat transfer coefficient related to the outside diameter can be calculated with Equation (2.27).

$$k_2 = \left(\frac{d_2}{d_1} \cdot \frac{1}{\alpha_{f1}} + \frac{d_2}{2 \cdot \lambda} \cdot \ln\left(\frac{d_2}{d_1} \right) + \frac{1}{\alpha_{f2}} \right)^{-1} =$$

$$= \left(\frac{15}{10.4} \cdot \frac{1}{15000} + \frac{0.015}{2 \cdot 40} \cdot \ln\left(\frac{15}{10.4} \right) + \frac{1}{13000} \right)^{-1} = 4137 \ \frac{W}{m^2 \cdot K}$$

b) The overall heat transfer coefficient related to the inner diameter is calculated with Equation (2.28).

$$k_1 = \left(\frac{1}{\alpha_{f1}} + \frac{d_1}{2 \cdot \lambda} \cdot \ln\left(\frac{d_2}{d_1} \right) + \frac{d_1}{d_2} \cdot \frac{1}{\alpha_{f2}} \right)^{-1} =$$

$$= \left(\frac{1}{15000} + \frac{0.0104}{2 \cdot 40} \cdot \ln\left(\frac{15}{10,4} \right) + \frac{10.4}{15} \cdot \frac{1}{13000} \right)^{-1} = 5966 \ \frac{W}{m^2 \cdot K}$$

c) The tube wall heat transfer coefficient calculated with Equation (2.21) is:

$$\alpha_{Wa} = \frac{2 \cdot \lambda}{d_a \cdot \ln(d_a / d_i)} = \frac{2 \cdot 40 \cdot W/(m \cdot K)}{0.015 \cdot m \cdot \ln(15/10.4)} = 14562 \ \frac{W}{m^2 \cdot K}$$

The approximated value calculated with Equation (2.30):

$$\alpha_{Wa} = \frac{\lambda}{s} = \frac{40 \cdot W/(m \cdot K)}{0.0023 \cdot m} = 17391 \ \frac{W}{m^2 \cdot K}$$

The value calculated with Equation (2.30) is 19 % too large. Reason: The ratio wall thickness to diameter is relatively large in this case, thus considering the hollow cylinder as a wall is not correct.

Discussion

It is extremely important to define the surface area to which the heat transfer coefficient is related. In this example the overall heat transfer coefficient related to the inner wall is 44 % larger than that to the outer wall. Using the heat transfer coefficient related to the inner surface area by mistake as that related to the outer surface area, the heat exchanger would be 44 % too small in area and thus in rated power.

The calculation of the wall heat transfer coefficient with the Equation (2.30) results in too high values. For an outer diameter 10 % larger than the inner one, (2.30) results in an error of 5 %. The deviation in Equation (2.21) can be shown by expansion in series.

$$\alpha_{Wa} = \frac{2 \cdot \lambda}{d_a \cdot \ln[1/(1 - 2 \cdot s/d_a)]} = \frac{2 \cdot \lambda}{d_a \cdot [2 \cdot s/d_a + (2 \cdot s/d_a)^2 / 2 + (2 \cdot s/d_a)^3 / 3 + ...]}$$

In Equation (2.30) the expansion in series is stopped after the first term.

2.1.4   Hollow cylinder with multiple layers

In technology applications tubes (hollow cylinders) often consist of multiple layers. Examples are heat exchanger tubes with corrosion resistant inner tubes, tubes with an outside insulation layer, corrosion, fouling and oxide layers inside and outside.

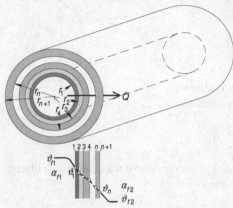

Figure 2.5: Thermal conduction in a hollow cylinder consisting of multiple layers

Figure 2.5 shows a hollow cylinder, with a wall consisting of n layers with different thicknesses and thermal conductivities. The heat transfer coefficients of the layers related to the outermost surface area are:

$$\alpha_i = \frac{2 \cdot \lambda_i}{d_{n+1} \cdot \ln(d_{i+1}/d_i)} \qquad (2.31)$$

The overall heat transfer coefficient is again the reciprocal of the sum of all relevant heat resistances:

$$\frac{1}{k} = \frac{d_{n+1}}{d_1} \cdot \frac{1}{\alpha_{f1}} + \sum_{i=1}^{i=n} \frac{d_{n+1}}{2 \cdot \lambda_i} \cdot \ln(d_{i+1}/d_i) + \frac{1}{\alpha_{f2}} \qquad (2.32)$$

To determine the wall temperatures the surface area change must be considered.

$$\frac{\vartheta_{f1} - \vartheta_1}{\vartheta_{f1} - \vartheta_{f2}} = \frac{d_{n+1}}{d_1} \cdot \frac{k}{\alpha_{f1}} \qquad \frac{\vartheta_i - \vartheta_{i+1}}{\vartheta_{f1} - \vartheta_{f2}} = \frac{d_{n+1}}{d_i} \cdot \frac{k}{\alpha_{Wa}} \qquad \frac{\vartheta_{n+1} - \vartheta_{f2}}{\vartheta_{f1} - \vartheta_{f2}} = \frac{k}{\alpha_{f2}} \qquad (2.33)$$

EXAMPLE 2.5: Condenser tube with fouling

In a titanium condenser tube of 24 mm outer diameter and 0.7 mm wall thickness an inside fouling layer of 0.05 mm is detected after a certain operation time. The thermal conductivity of titanium is 15 W/(m K), that of the fouling layer 0.8 W/(m K).
In the tube the heat transfer coefficient has the value of 18 000 W/(m² K), outside that of 13 000 W/(m² K).

Find

The reduction of the overall hat transfer coefficient.

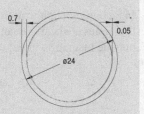

Solution

Schematics   See sketch.

Assumptions

- The thermal conductivity in the layers is constant.
- The wall temperatures of the layers are constant.

Analysis

The overall heat transfer coefficient related to the outermost wall can be calculated with Equation (2.32). For the clean tube it results in:

$$k = \left(\frac{d_2}{d_1} \cdot \frac{1}{\alpha_{f1}} + \frac{d_2}{2 \cdot \lambda_i} \cdot \ln(d_2 / d_i) + \frac{1}{\alpha_{f2}} \right)^{-1} =$$

$$= \left(\frac{24}{22.6} \cdot \frac{1}{18\,000} + \frac{0.024}{2 \cdot 15} \cdot \ln\left(\frac{24}{22.6} \right) + \frac{1}{13\,000} \right)^{-1} = 5435 \ \frac{W}{m^2 \cdot K}$$

For the tube with fouling the result is:

$$k_V = \left(\frac{d_3}{d_1} \cdot \frac{1}{\alpha_{f1}} + \frac{d_3}{2 \cdot \lambda_1} \cdot \ln(d_2 / d_1) + \frac{d_3}{2 \cdot \lambda_2} \cdot \ln(d_3 / d_2) + \frac{1}{\alpha_{f2}} \right)^{-1} =$$

$$= \left(\frac{24}{22.5} \cdot \frac{1}{18\,000} + \frac{0.024}{2 \cdot 0.8} \cdot \ln\left(\frac{22.6}{22.5} \right) + \frac{0.024}{2 \cdot 15} \cdot \ln\left(\frac{24}{22.6} \right) + \frac{1}{13\,000} \right)^{-1} = 3987 \ \frac{W}{m^2 \cdot K}$$

Here it is possible to show, that a simplified model provides almost the same result. With the simplification, the reciprocal of the fouling layer heat transfer coefficient is added to the reciprocal of the clean overall heat transfer coefficient, the fouled overall heat transfer coefficient results in:

$$k_V = \left(\frac{1}{k} + \frac{d_3}{2 \cdot \lambda_1} \cdot \ln(d_2 / d_3) \right)^{-1} = \left(\frac{1}{5435} + \frac{0.024}{2 \cdot 0.8} \cdot \ln\left(\frac{22.6}{22.5} \right) \right)^{-1} = 3992 \ \frac{W}{m^2 \cdot K}$$

This result is due to the very thin fouling layer, compared to the inner diameter of the tube.

Discussion

It could be shown that even a very thin fouling layer results in considerable reduction of the overall heat transfer coefficient. The reduction calculated in this example is 27 %, this is a realistic value. But also in titanium tubes without a fouling the corrosion resistant oxide layer already reduces the overall heat transfer coefficient by 6 to 8 %.

In practice, the thickness of fouling layers cannot exactly be measured. Further the determination of the thermal conductivity is rather difficult as the layers can have different thermal conductivities in dry and wet condition. Values for fouling resistances R_v collected in numerous tests, are used to take into account the influence of fouling when calculating the fouled overall heat transfer coefficients k_v.

$$k_v = (1/k + R_v)^{-1}.$$

EXAMPLE 2.6: Insulation of a steam pipe

In a steel pipe of 100 mm internal diameter and 5 mm wall thickness, hot steam with a temperature of 400 °C flows. In the tube the heat transfer coefficient is 1000 W/(m² K). The pipe must be protected with an insulation layer and with a 0.5 mm thick aluminum shell. The thermal conductivity of steel is 47 W/(m K), of aluminum 220 W/(m K) and of the insulator 0.08 W/(m K). According to the security requirements the temperature on the outer surface shall no exceed 45 °C at 32 °C room temperature and 15 W/(m² K) outside heat transfer coefficient.

Find

The required thickness of the insulation, if the material is available with different thicknesses in 10 mm steps and check which simplification could made.

Solution

Schematics   See sketch.

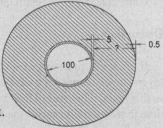

Assumptions

- The thermal conductivity of the layers is constant.
- The wall temperatures of the layers are constant.

Analysis

The temperature of outermost wall surface can be calculated with Equation (2.32). From there the overall heat transfer coefficient at which the outer surface temperature does not exceed the 45 °C can be determined.

$$k = \frac{\vartheta_{n+1} - \vartheta_{f2}}{\vartheta_{f1} - \vartheta_{f2}} \cdot \alpha_{f2} = \frac{45 - 32}{400 - 32} \cdot 15 \cdot \frac{W}{m^2 \cdot K} = 0.530 \ \frac{W}{m^2 \cdot K}$$

With Equation (2.32) the diameter of the insulation can be calculated.

$$\frac{1}{k} = \frac{d_4}{d_1} \cdot \frac{1}{\alpha_{f1}} + \frac{d_4}{2 \cdot \lambda_1} \cdot \ln\left(\frac{d_2}{d_1}\right) + \frac{d_4}{2 \cdot \lambda_2} \cdot \ln\left(\frac{d_3}{d_2}\right) + \frac{d_4}{2 \cdot \lambda_3} \cdot \ln(\frac{d_4}{d_3}) + \frac{1}{\alpha_{f2}}$$

As $d_4 = d_3 + 2s_3$ the unknown value is d_3. This equation cannot be solved analytically. Either an equation solver or an iteration delivers a result. The solution for d_3 is 294 mm. An outer diameter of **300 mm** was selected, with which an overall heat transfer coefficient of 0.511 W/(m² K) results.

The possible simplification can be demonstrated on the basis of the wall temperatures, calculated with Equation (2.33):

$$\vartheta_1 = \vartheta_{f1} - (\vartheta_{f1} - \vartheta_{f2})\frac{d_4}{d_1} \cdot \frac{k}{\alpha_{f1}} = 399.4 \ °C$$

$$\vartheta_2 = \vartheta_1 - (\vartheta_{f1} - \vartheta_{f2})\frac{d_4}{d_2} \cdot \frac{k}{\alpha_{w1}} = \vartheta_1 - (\vartheta_{f1} - \vartheta_{f2})\frac{d_4}{d_2} \cdot \frac{k \cdot d_4}{2 \cdot \lambda_1} \cdot \ln\left(\frac{d_2}{d_1}\right) = 399.3 \ °C$$

$$\vartheta_3 = \vartheta_2 - (\vartheta_{f1} - \vartheta_{f2})\frac{d_4}{d_3} \cdot \frac{k \cdot d_4}{2 \cdot \lambda_2} \cdot \ln\left(\frac{d_3}{d_2}\right) = 43.25 \ °C$$

$$\vartheta_4 = \vartheta_3 - (\vartheta_{f1} - \vartheta_{f2})\frac{k \cdot d_4}{2 \cdot \lambda_3} \cdot \ln\left(\frac{d_4}{d_3}\right) = 43.25 \ °C$$

The temperature drop in the metal walls and in the fluid is only 1.65 K. A calculation of the insulation layer with 400 °C inner and 45 °C outside temperature would provide an acceptable result. However, an iteration is still required.

Discussion

If in a wall consisting of several layers one of the layers has, compared to the other layers, a very low heat transfer coefficient, almost the total temperature drop occurs in this layer. The overall heat transfer coefficient is in this case only slightly smaller than that of the layer with the lowest heat transfer coefficient.

2.1.5   Thermal conduction in a hollow sphere

The determination of the heat rate in hollow spheres (Figure 2.6) is similar to that of the hollow cylinder. The surface area A for the heat flux changes with the radius adequate to the sphere surface area $A = 4 \cdot \pi \cdot r^2$.

$$\dot{Q} = -\lambda \cdot 4 \cdot \pi \cdot r^2 \cdot \frac{d\vartheta}{dr} \qquad (2.34)$$

The solution of this differential equation is:

$$\dot{Q} = \frac{\lambda \cdot 4 \cdot \pi}{1/r_1 - 1/r_2} \cdot (\vartheta_1 - \vartheta_2) = \frac{2 \cdot \lambda}{d_2 \cdot (d_2/d_1 - 1)} \cdot \pi \cdot d_2^2 \cdot (\vartheta_1 - \vartheta_2) \qquad (2.35)$$

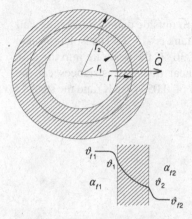

Figure 2.6: Thermal conductivity in a hollow sphere

The heat transfer coefficient related to the outer surface area is:

$$\alpha_{Wa} = \frac{2 \cdot \lambda}{d_2 \cdot (d_2/d_1 - 1)} \qquad (2.36)$$

The overall heat transfer coefficient of a hollow sphere that wall consists or multiple layers of different thicknesses and thermal conductivities is calculated as:

$$\frac{1}{k} = \frac{d_{n+1}^2}{d_1^2} \cdot \frac{1}{\alpha_{f1}} + \sum_{i=1}^{i=n} \frac{d_{n+1} \cdot (d_{n+1}/d_i - d_{n+1}/d_{i+1})}{2 \cdot \lambda_i} + \frac{1}{\alpha_{f2}} \qquad (2.37)$$

The wall temperatures of the layers can be calculated with Equation (2.38).

$$\frac{\vartheta_{f1} - \vartheta_1}{\vartheta_{f1} - \vartheta_{f2}} = \frac{d_{n+1}^2}{d_1^2} \cdot \frac{k}{\alpha_{f1}} \qquad \frac{\vartheta_i - \vartheta_{i+1}}{\vartheta_{f1} - \vartheta_{f2}} = \frac{d_{n+1}^2}{d_i^2} \cdot \frac{k}{\alpha_{Wa}} \qquad \frac{\vartheta_{n+1} - \vartheta_{f2}}{\vartheta_{f1} - \vartheta_{f2}} = \frac{k}{\alpha_{f2}} \qquad (2.38)$$

It is noteworthy that the heat rate of a hollow sphere with increasing wall thickness do not reach the value of zero. In plane and cylindrical walls the heat rate reaches zero with increasing wall thickness. The heat rate in a hollow sphere according to Equation (2.35) does not reach zero when the outer diameter goes to infinite.

$$\dot{Q} = \lim_{d_a \to \infty} \frac{2 \cdot \lambda}{(1/d_i - 1/d_a)} \cdot \pi \cdot (\vartheta_i - \vartheta_a) = 2 \cdot \lambda \cdot d_i \cdot \pi \cdot (\vartheta_i - \vartheta_a) \qquad (2.39)$$

More precisely, the heat transfer coefficient goes to zero but the surface area $A = \pi \cdot d^2$ goes to infinity and thus the heat flux has the value as given in Equation (2.39). This means that a sphere surface does always transfer heat to an infinite environment by heat conduction, as long as a temperature difference exists.

EXAMPLE 2.7: Insulation of a sphere-shaped tank

A spherical steel tank for carbon dioxide with 1.5 m outside diameter, 20 mm wall thickness shall be insulated. The temperature in the tank is –15 °C. The isolation shall be designed such that at an outside temperature of 30 °C the heat rate into the tank remains under 300 W. The inside and outside heat transfer resistances can be neglected. The insulation has a thermal conductivity of 0.05 W/(m K) and the steel 47 W/(m K).

Find

The required thickness of the insulation.

Solution

Assumptions

- The thermal conductivity of the tank wall and insulation is constant.
- The process is stationary.

Analysis

For not exceeding the given heat rate a corresponding small overall heat transfer coefficient is required. In this specific case it is useful to select the inner wall as reference surface, as the outer surface first after the determination of the insulation layer thickness will be known. The requested overall heat transfer coefficient related to the inner wall surface is:

$$k_{in} = \frac{\dot{Q}}{\pi \cdot d_1^2 \cdot (\vartheta_{f2} - \vartheta_{f1})} = \frac{300 \cdot \text{W}}{\pi \cdot 1.46^2 \cdot \text{m}^2 \cdot (30+15) \cdot \text{K}} = 0.996 \quad \frac{\text{W}}{\text{m}^2 \cdot \text{K}}$$

The neglecting of the fluid heat transfer resistances means, that the according heat transfer coefficients are assumed as infinite and inserted in Equation (2.36). With consideration of the relation of the heat transfer coefficient to the inner wall Equation (2.35) results in the following relations:

$$\dot{Q} = \frac{2 \cdot \lambda_1}{d_1 \cdot (1 - d_1 / d_2)} \cdot \pi \cdot d_1^2 \cdot (\vartheta_2 - \vartheta_1)$$

$$\dot{Q} = \frac{2 \cdot \lambda_2}{d_2 \cdot (1 - d_2/d_3)} \cdot \pi \cdot d_2^2 \cdot (\vartheta_3 - \vartheta_2) = \frac{d_2^2}{d_1^2} \cdot \frac{2 \cdot \lambda_2}{d_2 \cdot (1 - d_2/d_3)} \cdot \pi \cdot d_1^2 \cdot (\vartheta_3 - \vartheta_2)$$

With the two equations we receive:

$$\dot{Q} = \left(\frac{d_1 \cdot (1 - d_1/d_2)}{2 \cdot \lambda_1} + \frac{d_2 \cdot (1 - d_2/d_3)}{2 \cdot \lambda_2} \cdot \frac{d_1^2}{d_2^2} \right)^{-1} \cdot \pi \cdot d_1^2 \cdot (\vartheta_3 - \vartheta_1)$$

The overall heat transfer coefficient is the term in the brackets in the above equation. Its value is known and the equation can be solved for d_3.

$$d_3 = \frac{d_2}{1 - \left(\frac{1}{k} - \frac{d_1 \cdot (1 - d_1/d_2)}{2 \cdot \lambda_1} \right) \cdot \frac{2 \cdot \lambda_2 \cdot d_2}{d_1^2}} =$$

$$= \frac{1.5 \cdot m}{1 - \left(\frac{1}{0.996} - \frac{1.46 \cdot (1 - 1.46/1.5)}{2 \cdot 47} \right) \frac{2 \cdot 0.05 \cdot 1.5}{1.46^2}} = 1.612 \ m$$

Discussion

For some specific problems it is convenient not to relate the heat transfer coefficients to the outer surface area. If not all required temperatures are given, connecting the rate equations for different layers can deliver the required solution.

2.1.6 Thermal conduction with heat flux to extended surfaces

To extend the heat transfer surface area of heat exchangers *fins* can be installed. Also support rods and feet can transfer heat to or from tanks. Fins discussed here have a constant cross-section. If the side walls would be ideally insulated we would have the case of thermal conduction in a plane wall, in which a linear temperature gradient would develop, constant thermal conductivity provided. The fins and rods discussed here have no insulated side walls, thus they are transferring heat form or to their environment. The heat rate in the fins is not constant. It is changing according to the transferred heat through the side walls. This is a two-dimensional phenomenon. A plane wall with constant cross-section, kept to a constant temperature ϑ_0 at the base (fin foot) and with heat transferred through the side surfaces can be handled as one-dimensional problem if the temperature in each cross sectional area can be assumed as constant. Figure 2.7 shows a fin with constant square cross-section installed on a surface with a constant temperature ϑ_0, from where a constant heat rate enters the fin.

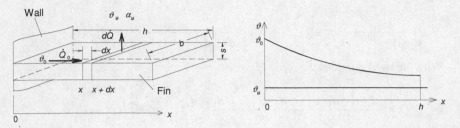

Figure 2.7: Thermal conduction in a fin

The fin is in contact with a fluid of lower temperature ϑ_u and transfers heat to the fluid. The heat transfer coefficient in the fluid is α_u.

Now the task is to determine the temperature distribution and heat rate in the fin.

2.1.6.1 Temperature distribution in the fin

The heat rate at the location x that enters the volume element $b \cdot s \cdot dx$ through the surface area $A = s \cdot b$ is:

$$\dot{Q}_x = -\lambda \cdot A \cdot \frac{d\vartheta}{dx} \tag{2.40}$$

At the location $x + dx$ the following heat rate leaves the volume element:

$$\dot{Q}_{x+dx} = \dot{Q}_x + \frac{\delta\dot{Q}_x}{dx} \cdot dx = -\lambda \cdot A \cdot \frac{d\vartheta}{dx} - \lambda \cdot A \cdot \frac{d^2\vartheta}{dx^2} \cdot dx \tag{2.41}$$

The change in the heat rate is:

$$\delta\dot{Q}_x = \dot{Q}_{x+dx} - \dot{Q}_x = -\lambda \cdot A \cdot \frac{d^2\vartheta}{dx^2} \cdot dx \tag{2.42}$$

This change in heat rate must equal the heat rate which leaves the volume element at its side walls according to the first law of thermodynamics. The heat rate through the side wall element $U \cdot dx$ is determined by the heat transfer coefficient and temperature difference.

$$-\delta\dot{Q}_x = \alpha_U \cdot U \cdot (\vartheta - \vartheta_U) \cdot dx \tag{2.43}$$

The circumference of the fin is U. Equations (2.42) and (2.43) deliver the following differential equation:

$$\frac{d^2\vartheta}{dx^2} = \frac{\alpha_U \cdot U}{\lambda \cdot A} \cdot (\vartheta - \vartheta_U) \tag{2.44}$$

Assuming a constant outside temperature ϑ_u, constant heat transfer coefficient α_u and constant thermal conductivity λ, the first term on the right side of Equation (2.44) can be replaced by the constant m^2. With substituting $\vartheta - \vartheta_U$ by $\Delta\vartheta$, the following differential equation results:

$$\frac{d^2 \Delta\vartheta}{dx^2} = m^2 \cdot \Delta\vartheta \tag{2.45}$$

$$m = \sqrt{\frac{\alpha_U \cdot U}{\lambda \cdot A}} \quad \text{and} \quad \Delta\vartheta = (\vartheta - \vartheta_U)$$

Solving the differential equation results in:

$$\Delta\vartheta = C_1 \cdot e^{-m \cdot x} + C_2 \cdot e^{m \cdot x} \tag{2.46}$$

The constants C_1 and C_2 must be determined with the boundary conditions. The boundary conditions can be obtained on both ends of the fin.

At $x = 0$ the temperature difference is $\Delta\vartheta_0$.

If the heat rate at the end of fin may be neglected i.e. at $x = h$ the heat rate is zero, the temperature gradient has the value $(d\vartheta / dx) = 0$.

With these two boundary conditions we receive:

$$\Delta\vartheta_0 = C_1 + C_2 \tag{2.47}$$

$$\left(\frac{\Delta\vartheta}{dx}\right)_{x=h} = -m \cdot C_1 \cdot e^{-m \cdot h} + m \cdot C_2 \cdot e^{m \cdot h} = 0 \tag{2.48}$$

From Equation (2.48) the relationship between C_1 and C_2 is:

$$C_1 = C_2 \cdot \frac{e^{m \cdot h}}{e^{-m \cdot h}} \tag{2.49}$$

Equation (2.49) inserted in Equation (2.47), delivers for C_2:

$$C_2 = \Delta\vartheta_0 \cdot \frac{e^{-m \cdot h}}{e^{m \cdot h} + e^{-m \cdot h}} \tag{2.50}$$

Equation (2.50) inserted in Equation (2.49), determines C_1:

$$C_1 = \Delta\vartheta_0 \cdot \frac{e^{+m \cdot h}}{e^{m \cdot h} + e^{-m \cdot h}} \tag{2.51}$$

We receive for the temperature difference $\Delta\vartheta$:

$$\Delta\vartheta(x) = \Delta\vartheta_0 \cdot \frac{e^{-m \cdot (h-x)} + e^{m \cdot (h-x)}}{e^{m \cdot h} + e^{-m \cdot h}} \tag{2.52}$$

For an *infinitely long fin* the negative exponential functions have the value of zero and Equation (2.51) delivers:

$$\Delta\vartheta(x) = \Delta\vartheta_0 \cdot e^{-m \cdot x} \tag{2.53}$$

For finite long fins in Equation (2.52) the exponential functions can be replaced by hyperbola functions.

$$\frac{\Delta\vartheta(x)}{\Delta\vartheta_0} = \frac{\cosh[m\cdot(h-x)]}{\cosh(m\cdot h)} = \frac{\cosh[m\cdot h\cdot(1-x/h)]}{\cosh(m\cdot h)} \qquad (2.54)$$

Figure 2.8 shows the relative temperature $\Delta\vartheta/\Delta\vartheta_0$ versus the relative length x/h with characteristic term of fins $m\cdot h$ as parameter.

For large values of $m\cdot h$ the temperature of the fin changes rapidly. In fins with low heat conductivity and large outside heat transfer coefficients, a large change of temperature occurs.

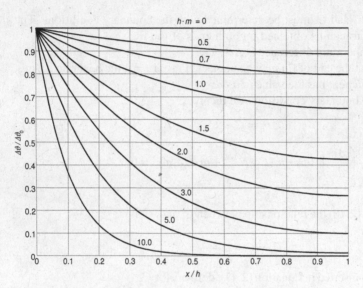

Figure 2.8: Relative temperature versus relative length

2.1.6.2   *Temperature at the fin tip*

The temperature at the fin tip is:

$$\vartheta(h) = \vartheta_U + \Delta\vartheta(h) = \vartheta_U + \Delta\vartheta_0 \cdot \frac{1}{\cosh(m\cdot h)} \qquad (2.55)$$

2.1.6.3   *Heat rate at the fin foot*

One of most important terms is the heat rate at the fin foot, because it is the total heat rate transferred from or to the fin. We determine it with Equation (2.40), using the spacial temperature gradient in the fin at $x = 0$.

$$\dot{Q}_{x=0} = -\lambda\cdot A\cdot\left(\frac{d\vartheta}{dx}\right)_{x=0} = -\lambda\cdot A\cdot\Delta\vartheta_0\cdot m\cdot\frac{-\sinh(m\cdot h)}{\cosh(m\cdot h)} =$$
$$= \lambda\cdot A\cdot\Delta\vartheta_0\cdot m\cdot\tanh(m\cdot h) \qquad (2.56)$$

As the heat rate at the fin foot is the same as that transferred through the total fin surface, it can also be determined by using the temperature difference from Equation (2.54) in Equation (2.43) and integrating the resulting equation from $x = 0$ to $x = h$.

2.1.6.4 Fin efficiency

Fins are installed to extend the heat transfer surface area. To have an effective extension of the surface area, the temperature in the fin should only be subject to a small temperature change. As the fin reaches the temperature of the surroundings the heat rate decreases, thus the material is not used as efficiently as closer to the base. The temperature in an ideal fin would not change, i.e. the heat transfer would occur always at the largest temperature difference $\Delta\vartheta_0$. In this ideal case the heat rate would be:

$$\dot{Q}_{ideal} = U \cdot h \cdot \alpha_U \cdot \Delta\vartheta_0 \qquad (2.57)$$

The ratio of the real heat rate to the ideal one, according to Equation (2.57), is the *fin efficiency* η_{Ri}. The real heat rate is given by Equation (2.56).

$$\eta_{Ri} = \frac{\dot{Q}_{x=0}}{\dot{Q}_{ideal}} = \frac{\lambda \cdot A \cdot m}{U \cdot \alpha_U \cdot h} \cdot \tanh(m \cdot h) = \frac{\tanh(m \cdot h)}{m \cdot h} \qquad (2.58)$$

Figure 2.9 demonstrates the fin efficiency versus characteristic fin term $m \cdot h$.

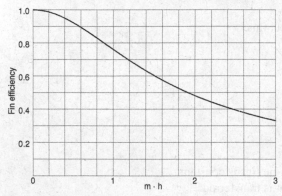

Figure 2.9: Fin efficiency versus $m \cdot h$

As already shown in Figure 2.8, the fin temperatures underlie a larger change with higher $m \cdot h$. This is the reason for the dropping of the fin efficiency.

> *The fin efficiency decreases with increasing fin height h and heat transfer coefficient α_U and circumference to cross-section ratio, with increasing thermal conductivity of the fin, its efficiency is also increasing.*

Fins are economical, if the additional costs for the fins result in lower total cost of the heat exchanger. As a rule of thumb, the fin efficiency should be larger than 0.8.

Short fins with high conductivity and high ratio of circumference to cross-section leads to high fin efficiency. Finned surfaces mainly are selected at low heat transfer coefficients, witch are familiar to gases.

2.1.6.5 Applicability for other geometries

Here the rules were developed for a square shaped fin with constant cross section. They are applicable without any restriction for fins with a constant cross section e.g. round rods, T-rods etc. Finned tubes will be discussed in Chapter 3.3.

EXAMPLE 2.8: Enlargement of the heat exchanger surface area by fins

A boiler has plane steel walls. To extend the surface area cylindrical steel fins of the same material with 25 mm height and 8 mm diameter are installed. The fins are welded to the steel plate and have a squared arrangement with 8 mm gaps between the fins. The heat transfer coefficient of the flue gas is 50 W/(m² K). The thermal conductivity of the fins is 17 W/(m K). The temperature of the wall is 100 °C, that of the flue gas 1000 °C.

Find

a) Extension of the surface area.
b) The heat flux with and without fins.
c) The temperature at the fin tip.

Solution

Schematics   See sketch.

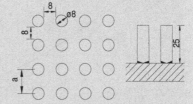

Assumptions

- The thermal conductivity in the fins is constant.
- At the end of the fins no heat is transferred.
- The temperature in the fins changes only in axial direction.
- The fins have metallic contact with the steel plate.

Analysis

a)   For the installation of one fin a surface area of square shape, with a side length from fin center to fin center, which is 16 mm, is used. Per fin, a surface area of 256 mm² is required, thus 3906 fins can be installed on a surface area of 1 m². The total heat transfer surface area is the surface area A of the plate reduced by the cross-sections of the fins A_0 plus the surface area of the fins A_{Ri}. The specific increase of the surface area is:

$$\frac{A_{tot}}{A} = \frac{A - A_0 + A_{Ri}}{A} = 1 - \frac{\pi}{4} \cdot \frac{d_{Ri}^2}{a^2} + \pi \cdot \frac{d_{Ri} \cdot h}{a^2} = 3.258$$

b) The heat flux to the plate without the fins is:

$$\dot{q}_{without} = \alpha_U \cdot (\vartheta_U - \vartheta_0) = 50 \cdot \frac{W}{m^2 \cdot K} \cdot (1000 - 100) \cdot K = 45 \ \frac{kW}{m^2}$$

With the installation of the fins the surface area of the plate is reduced, but additional heat will be transferred by the fins. The heat rate transferred by the fins can either be calculated with Equation (2.56) or with using the fin efficiency. Both methods will be shown.

$$\dot{q}_{with} = \frac{A - A_0}{A} \alpha_U \cdot (\vartheta_U - \vartheta_0) + \frac{1}{a^2} \cdot \dot{Q}_0 =$$

$$= (1 - \frac{\pi}{4} \cdot \frac{d_{Ri}^2}{a^2}) \cdot \alpha_U \cdot (\vartheta_U - \vartheta_0) + \frac{\lambda \cdot A_{QRi} \cdot \Delta\vartheta_0 \cdot m}{a^2} \cdot \tanh(m \cdot h)$$

$$\dot{q}_{with} = \frac{A - A_0 + A_{Ri} \cdot \eta_{Ri}}{A} \cdot \alpha_U \cdot (\vartheta_U - \vartheta_0) = \left(1 - \frac{\pi}{4} \cdot \frac{d_{Ri}^2}{a^2} + \pi \cdot \frac{d_{Ri} \cdot h}{a^2} \cdot \eta_{Ri}\right) \cdot \alpha_U \cdot (\vartheta_U - \vartheta_0)$$

For both equations the value of m is required.

$$m = \sqrt{\frac{\alpha_U \cdot U}{\lambda \cdot A_{QRi}}} = \sqrt{\frac{\alpha_U \cdot 4 \cdot \pi \cdot d_{Ri}}{\lambda \cdot \pi \cdot d_{Ri}^2}} = \sqrt{\frac{\alpha_U \cdot 4}{\lambda \cdot d_{Ri}}} = \sqrt{\frac{50 \cdot 4}{17 \cdot 0.008 \cdot m^2}} = 38.35 \ m^{-1}$$

Fin efficiency: $\qquad \eta_{Ri} = \frac{\tanh(m \cdot h)}{m \cdot h} = 0.776$

The upper equation gives the heat rate as:

$$\dot{q}_{wih} = [\alpha_U - 0.25 \cdot \pi \cdot d_{Ri}^2 / a^2 \cdot (1 - \lambda \cdot m \cdot \tanh(m \cdot h))] \cdot (\vartheta_U - \vartheta_0) =$$

$$= [50 - 3906 \cdot \pi / 4 \cdot 0.008^2 \cdot (50 - 17 \cdot 38.35 \cdot \tanh(38.35 \cdot 0.025))] \cdot 900 = 121.8 \ \frac{kW}{m^2}$$

The lower equation delivers:

$$\dot{q}_{with} = (1 - n \cdot \pi \cdot d_{Ri} \cdot (d_{Ri} / 4 - h \cdot \eta_{Ri}) \cdot 45 \cdot \frac{kW}{m^2} =$$

$$= (1 - 3906 \cdot \pi \cdot 0.008 \cdot (0.002 - 0.025 \cdot 0.776) \cdot 45 \cdot \frac{kW}{m^2} = 121.8 \ \frac{kW}{m^2}$$

The ratio of heat flux with and without fins is:

$$\dot{q}_{mit} / \dot{q}_{ohne} = 2.708$$

c) With Equation (2.51) the fin tip temperature can be determined.

$$\vartheta(h) = \vartheta_U + (\vartheta_0 - \vartheta_U) \cdot \frac{1}{\cosh(m \cdot h)} = 1000 \ °C - \frac{900 \cdot K}{\cosh(38.35 \cdot 0.025)} = \mathbf{398 \ °C}$$

Discussion

By installing the fins the heat transfer surface area is increased by a factor of 3.258, the heat flux by a factor of 2.7. The non-proportional increase of the heat flux is due to the temperature change in the fins. Consequently, the heat flux decreases along the fins. To determine the heat flux, two different calculation methods were demonstrated, both yielding the same result.

EXAMPLE 2.9: Heat transfer through a fixation rod

A tank for hot steam is fixed with rectangular steel rods of 20 mm x 40 mm. The rod's length is 400 mm. The tank is coated with a 100 mm insulator, e.g. the first 100 mm of the rod is within the insulation. It can be assumed that this heat insulation is ideal. The ambient heat transfer coefficient of the air, outside the insulation, has the value of 5 W/(m² K). The tank surface has a temperature of 150 °C, the air 20 °C. The thermal conductivity of the rod is 47 W/(m K).

Find:

a) Temperature distribution along the rod.
b) Insulation thickness, to avoid rod temperatures outside the insulation exceeding 90 °C.

Solution

Schematics   See sketch.

Assumptions

- The thermal conductivity in the rod is constant.
- At the fin tip no heat transfer occurs.
- In the rod the temperature changes only in axial direction.
- Between rod and tank there is a metallic contact.

Analysis

a) From the part of the rod (the first 100 mm) which is surrounded by the insulation, no heat is transferred through the side walls. Therefore, this part can be treated as a plane wall with the thickness s, the remaining part of the rod is a fin with the height h. According to Equation (2.6) the heat rate in the plane wall is:

$$\dot{Q} = \frac{\lambda}{s} \cdot A \cdot (\vartheta_W - \vartheta_0)$$

This heat rate enters the fin and its value can be calculated with Equation (2.56).

$$\dot{Q} = \lambda \cdot A \cdot \Delta\vartheta_0 \cdot m \cdot \tanh(m \cdot h) = \lambda \cdot A \cdot (\vartheta_0 - \vartheta_U) \cdot m \cdot \tanh(m \cdot h)$$

The temperature ϑ_0 at the beginning of the fin is not yet known. From both equations ϑ_0 can be determined.

$$\vartheta_0 = \frac{\vartheta_W + \vartheta_U \cdot s \cdot m \cdot \tanh(m \cdot h)}{s \cdot m \cdot \tanh(m \cdot h) + 1}$$

First the value of m must be calculated.

$$m = \sqrt{\frac{\alpha_U \cdot U}{\lambda \cdot A}} = \sqrt{\frac{\alpha_U \cdot 2 \cdot (a+b)}{\lambda \cdot a \cdot b}} = \sqrt{\frac{10 \cdot 2 \cdot (0.02 + 0.04)}{47 \cdot 0.02 \cdot 0.04 \cdot m^2}} = 5.649 \ \text{m}^{-1}$$

The temperature ϑ_0 we receive is:

$$\vartheta_0 = \frac{150 \cdot {}^{\circ}C + 20 \cdot {}^{\circ}C \cdot 0.1 \cdot m \cdot 5.649 \cdot m^{-1} \cdot \tanh(5.649 \cdot 0,3)}{0,1 \cdot m \cdot 5.649 \cdot m^{-1} \cdot \tanh(5.649 \cdot 0.3) + 1} = 105.07 \ {}^{\circ}C$$

In the first 100 mm of the rod the temperature drops linearly from 150 to 105.07 °C. Outside of the insulation the temperature along the rod can be determined by Equation (2.54):

$$\vartheta(x) = \vartheta_U + (\vartheta_0 - \vartheta_U) \cdot \frac{\cosh[m \cdot (h-x)]}{\cosh(m \cdot h)}$$

The following values were calculated:

x	$\vartheta(x)$
m	°C
0.00	150.00
0.10	105.07
0.13	92.76
0.16	82.53
0.19	74.11
0.22	67.25
0.25	61.74
0.28	57.44
0.31	54.22
0.34	51.98
0.37	50.66
0.40	50.23

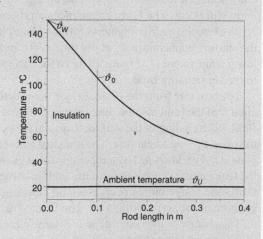

The diagram shows the temperature distribution along the rod.

b) To determine the insulation thickness required to not exceed a rod surface temperature of 90 °C outside the insulation, the equation used for the calculation of the temperature at the beginning of the fin can be used. The value of the length s is the unknown quantity. The height h of the fin, which is now $h = l - s$, will be inserted.

$$\vartheta_0 = \frac{\vartheta_W + \vartheta_U \cdot s \cdot m \cdot \tanh[m \cdot (l - s)]}{s \cdot m \cdot \tanh[m \cdot (l - s)] + 1}$$

This equation cannot be solved analytically, but it converges fast. A simple iteration procedure results in: **179 mm**.

Discussion

The heat rate transferred outside the insulation from the rod determines the temperature change in the insulated part. Outside the insulation there is a rather slow change in temperature. At the end of the rod the temperature is 55.2 °C, i.e. 35 K higher than the ambient temperature.

2.2   Transient thermal conduction

2.2.1   One-dimensional transient thermal conduction

2.2.1.1   Determination of the temporal change of temperature

Putting a body with the starting temperature ϑ_A into an environment with another temperature, a temporal and spacial change of the temperature distribution in the body will occur. The body experiences a process of *transient thermal conduction*. As an example, we investigate an infinite plane plate with the thickness $2\,s$, which has the starting temperature ϑ_A at the time $t = 0$, and will be brought into a fluid with a lower temperature ϑ_∞ (Figure 2.10). At the surface of the plate heat will be transferred to the surrounding fluid.

The heat rate from the plate surface are is determined by the heat transfer coefficient α of the fluid and the temperature difference between the plate surface and the fluid. With cooling of the surface a spacial temperature gradient will be generated in the plate, driving a heat flux. The transient and spacial temperature distribution in the plate is not yet known. The temperature change shown in Figure 2.10 shows only that the temperature decreases from the middle to the surface of the plate and that - with time - the temperatures decreases and converges asymptotically towards the temperature of the surroundings ϑ_∞. The latter is reached at infinity only.

The change of the heat rate in a volume element of the plate is due to the change of the spacial temperature gradient. The heat rate causes a decrease of the enthalpy (heat content) and consequently that of the temperature. The tangent of the temperature gradient at the plate surface matches the temperature ϑ_∞ in a distance of λ/α.

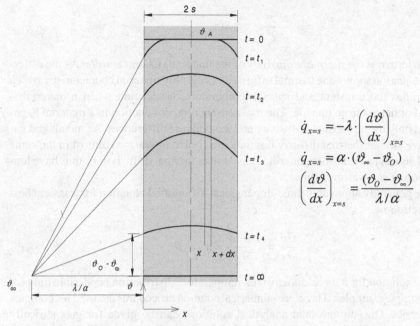

$$\dot{q}_{x=s} = -\lambda \cdot \left(\frac{d\vartheta}{dx}\right)_{x=s}$$

$$\dot{q}_{x=s} = \alpha \cdot (\vartheta_\infty - \vartheta_0)$$

$$\left(\frac{d\vartheta}{dx}\right)_{x=s} = \frac{(\vartheta_0 - \vartheta_\infty)}{\lambda/\alpha}$$

Figure 2.10: Temperature development due to outside cooling of a plane plate

As the side dimensions of the plate are infinite, only heat transfer in x direction is possible. The heat rate to a volume element in x direction can be defined by the following differential equation:

$$\dot{Q}_x = -\lambda \cdot A \cdot \frac{\partial \vartheta}{\partial x} \tag{2.59}$$

At the location $x + dx$ the heat rate is:

$$\dot{Q}_{x+dx} = \dot{Q}_x + \frac{\partial \dot{Q}_x}{\partial x} \cdot dx = -\lambda \cdot A \cdot \frac{\partial \vartheta}{\partial x} - \lambda \cdot A \cdot \frac{\partial^2 \vartheta}{\partial x^2} \cdot dx \tag{2.60}$$

We receive for the change of the heat rate:

$$d\dot{Q}_x = \dot{Q}_{x+dx} - \dot{Q}_x = -\lambda \cdot A \cdot \frac{\partial^2 \vartheta}{\partial x^2} \cdot dx \tag{2.61}$$

As only heat transfer in x direction is possible, the transient change of the heat rate is equivalent to the change of the enthalpy of the volume element, which is defined as:

$$\delta \dot{Q}_x = -\rho \cdot A \cdot dx \cdot c_p \cdot \frac{\partial \vartheta}{\partial t} \tag{2.62}$$

With Equation (2.61) and Equation (2.62) we receive the differential equation for the temporal and spacial temperature distribution in the plate.

$$\frac{\partial \vartheta}{\partial t} = a \cdot \frac{\partial^2 \vartheta}{\partial x^2} \qquad \text{with} \qquad a = \frac{\lambda}{\rho \cdot c_p} \qquad (2.63)$$

The term a is the *thermal diffusivity* of the material. Its unit is **m²/s**. As the differential equation shows, the thermal diffusivity is the only material characteristic which determines the transient and spacial temperature distribution when a material is cooled or heated from outside. The transient thermal conduction in a material is governed only by the thermal diffusivity and temperature differences. As metals and gases have similar thermal diffusivities they are cooled or warmed almost in the same time. Liquids and non-metallic solids have lower thermal diffusivities, thus have longer cooling and warming times.

The generally applicable three-dimensional differential equation for transient heat conduction is:

$$\frac{\partial \vartheta}{\partial t} = a \cdot \frac{\partial^2 \vartheta}{\partial r^2} = a \cdot \nabla^2 \vartheta \qquad (2.64)$$

The solution for a three-dimensional temperature distribution is possible only for a few simple examples. Therefore, numerical solution procedures are used for complex geometries. One-dimensional analytical solutions can be given for geometrically simple problems in the form of so called *Fourier series*. For the one-dimensional plate one generally applicable solution is:

$$\vartheta(x,t) = \sum_{n=1}^{\infty} \left[C_1 \cdot \cos(B \cdot x) + C_2 \cdot \sin(B \cdot x) \right] \cdot e^{-a \cdot C_3 \cdot t} \qquad (2.65)$$

For the corresponding boundary conditions the constants B and C can be determined. In the case of infinitely large heat transfer coefficients in the surroundings, i.e. the surface of the plate having the temperature of the surroundings, we receive the following solution:

$$\frac{\vartheta - \vartheta_\infty}{\vartheta_A - \vartheta_\infty} = \frac{4}{\pi} \cdot \sum_{n=1}^{\infty} \frac{1}{n} \cdot e^{-\frac{n^2 \cdot \pi^2 \cdot a \cdot t}{4 \cdot s^2}} \cdot \sin \frac{n \cdot \pi \cdot x}{2 \cdot s} \quad \text{with} \quad n = 1,3,5... \qquad (2.66)$$

Fourier series show a rapid convergence but the calculation of the series is time-consuming. In Figures 2.11 to 2.13 solutions for a plane plate, cylinder and sphere are presented. With the diagrams, local temperatures at the surface and in the center as well as the caloric average temperatures of the body can be determined.

In the diagrams the following dimensionless characteristics are used:

dimensionless temperature Θ:
$$\Theta = \frac{\vartheta - \vartheta_\infty}{\vartheta_A - \vartheta_\infty} \qquad (2.67)$$

Fourier number Fo:
$$Fo = a \cdot t / s^2 \qquad (2.68)$$

Biot number Bi: $\qquad\qquad\qquad\qquad Bi = \alpha \cdot s / \lambda \qquad\qquad\qquad$ (2.69)

The dimensionless temperature is the ratio of the difference between local and ambient temperature $\vartheta - \vartheta_\infty$ to the largest temperature difference in the process, which is $\vartheta_A - \vartheta_\infty$.

The *Fourier* number is a *dimensionless time*. It is the ratio of heat rate to the temporal change of the enthalpy.

The *Biot* number is the ratio of the conduction heat transfer resistance in the body and the convection heat transfer resistance between body and surrounding liquid.

In the diagrams 2.11 to 2.13 the dimensionless temperature is given as a function of the *Fourier* and *Biot* number. The index O is for the surface temperatures of the body and m stands for the center of the body. The dimensionless caloric medium temperature $\overline{\Theta}$ is the integrated mean temperature of the body. In the diagrams, X is half of the plate thickness, R is the radius of the cylinder and sphere.

The temperature reached after a certain time can be determined with the *Fourier* and *Biot* number. The time required for a certain temperature change can be calculated with the *Fourier* number, received with the dimensionless temperature and *Biot* number. Heat transfer coefficients, required for a certain temperature change in a given time, can be determined with the *Biot* number, obtained with the dimensions-less temperature and the *Fourier* number from the diagrams.

2.2.1.2 Determination of transferred heat

With the mean temperature the transferred heat in a given time to or from a body can be determined:

$$Q = H - \overline{H} = m \cdot (h_A - \overline{h}) = m \cdot c_p \cdot (\vartheta_A - \overline{\vartheta}) \qquad\qquad (2.70)$$

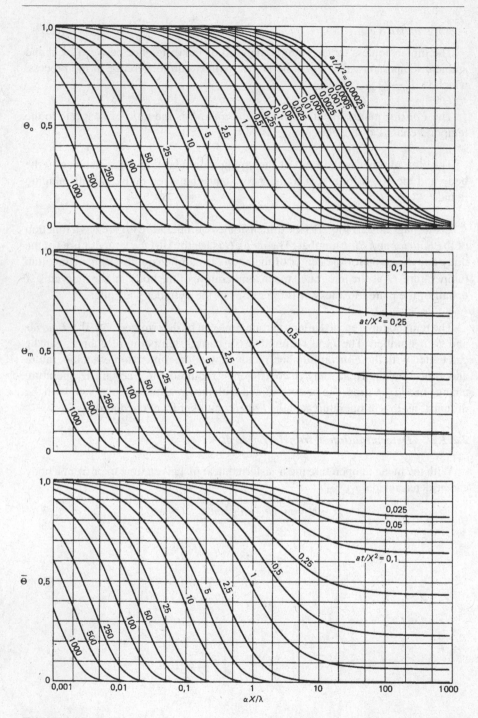

Figure 2.11: Temperature development of transient thermal conduction in a plate [2.1]

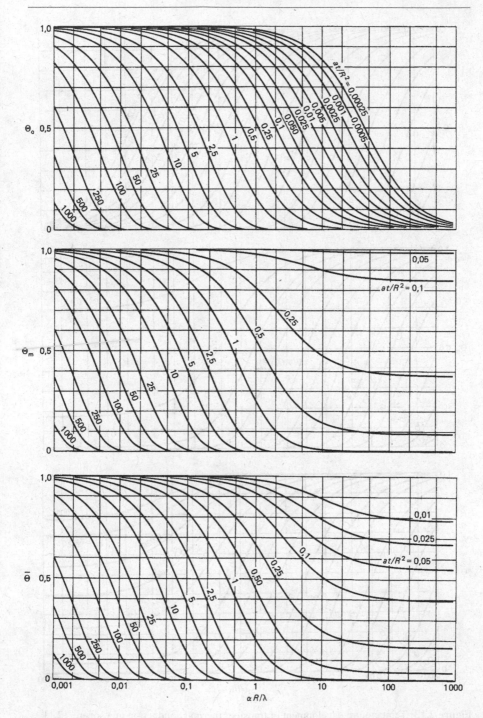

Figure 2.12: Temperature development of transient thermal conduction in a circular cylinder [2.1]

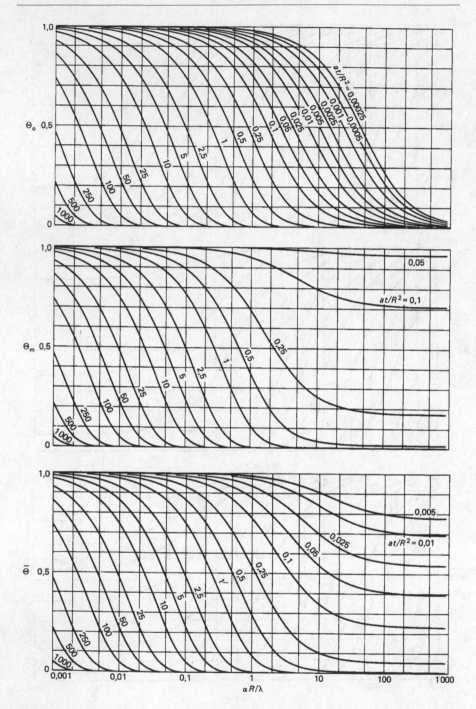

Figure 2.13: Temperature development of transient thermal conduction in a sphere [2.1]

EXAMPLE 2.10: Cooling of a plastic sheet

A plastic sheet of 4 mm thickness and 1 m width leaves a roll with a temperature of 150 °C. It is cooled with an air fan. After some distance it is cut into pieces of 2 m length. To avoid plastic deformation during the cutting process, the temperature in the middle of the sheet must be below 50 °C. The air temperature is 25 °C and the heat transfer coefficient 50 W/(m² K). The material properties of the plastic are:

$$\rho = 2400 \text{ kg/m}^3, \ \lambda = 0.8 \text{ W/(m K)} \ c_p = 800 \text{ J/(kg K)}.$$

Find

a) The speed of the plastic sheet.
b) Which heat rate is removed from the sheet.

Solution

Schematic   See sketch.

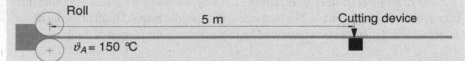

Roll   5 m   Cutting device

$\vartheta_A = 150 \text{ °C}$

Assumptions

- The material properties in the sheet are constant.
- Heat transfer from the side can be neglected.
- The speed of the sheet is constant.

Analysis

a)   With the time required to cool the sheet from 150 °C to 50 °C the speed can be determined. The time can be determined using the *Fourier* number. As it is a function of the dimensionless temperature and the *Biot* number, first these two parameters must be calculated. The dimensionless temperature according to Equation (2.67) is:

$$\Theta_m = \frac{\vartheta_m - \vartheta_\infty}{\vartheta_A - \vartheta_\infty} = \frac{50 - 25}{150 - 25} = 0.2$$

The *Biot* number is determined with Equation (2.69):

$$Bi = \frac{\alpha \cdot s}{\lambda} = \frac{50 \cdot \text{W} \cdot 0.002 \cdot \text{m} \cdot \text{m} \cdot \text{K}}{\text{m}^2 \cdot \text{K} \cdot 0.8 \cdot \text{W}} = 0.125$$

The central diagram in Figure 2.11 delivers the *Fourier* number value $Fo = 15$. To calculate the required time, first the thermal diffusivity must be determined.

$$a = \frac{\lambda}{\rho \cdot c_p} = \frac{0.8 \cdot W \cdot m^3 \cdot kg \cdot K}{m \cdot K \cdot 2400 \cdot kg \cdot 800 \cdot J} = 4.17 \cdot 10^{-7} \frac{m^2}{s}$$

With the *Fourier* number and Equation (2.68) we find for the required time:

$$t = \frac{Fo \cdot s^2}{a} = \frac{15 \cdot 0.002^2 \cdot m^2 \cdot s}{4.17 \cdot 10^{-7} \cdot m^2} = 144 \text{ s}$$

This time is required to cool the sheet when traveling the 5 m distance to the cutter. The velocity is $x/t = $ **0.0347 m/s = 2.08 m/min**.

b) To determine the heat rate, first the specific heat removed per kg from the sheet must be known. This can be calculated with Equation (2.70). For this calculation from the bottom diagram in Figure 2.11 the dimensionless mean temperature $\overline{\Theta}$ can be estimated. With the *Fourier* number of 15 and *Biot* number 0.125 we receive: $\overline{\Theta} = 0.17$. The mean temperature of the sheet after 5 m is:

$$\overline{\vartheta} = \vartheta_\infty + (\vartheta_A - \vartheta_\infty) \cdot \overline{\Theta} = 25 \text{ °C} + (150 - 25) \cdot K \cdot 0.17 = 46.25 \text{ °C}$$

After dividing both sides of Equation (2.70) by the mass, we receive the specific heat q, removed per kg mass. The heat rate is the specific heat multiplied by the mass flow rate. The latter can be determined by the well-known equation of fluid mechanics.

$$\dot{m} = c \cdot \rho \cdot 2 \cdot s \cdot b = 0.0347 \cdot m/s \cdot 2400 \cdot kg/m^3 \cdot 2 \cdot 0.002 \cdot m \cdot 1 \cdot m = 0.333 \text{ kg/s}$$

We receive for the heat rate:

$$\dot{Q} = \dot{m} \cdot q = \dot{m} \cdot c_p \cdot (\vartheta_A - \overline{\vartheta}) = 0.417 \cdot kg/s \cdot 800 \cdot J/(kg \cdot K) \cdot (150 - 46.25) \cdot K = \textbf{27.67 kW}$$

Discussion

Many technical problems can easily be calculated with the diagrams presented in the book. However, more effects often have to be considered. In our example the air would be blown in counterflow to the motion of the sheet and its temperature would not remain constant. Taking into account the temperature rise of the air, a step-by-step calculation could be performed. However, a computer code would be required for such a procedure. Our calculation is valid only when the temperature rise of the air is not too large.

EXAMPLE 2.11: Cooling of beer cans

For a barbecue, beer cans should be cooled from 30 °C to 4 °C mean temperature in a refrigerator. The cans have a diameter of 65 mm. The temperature in the refrigerator has the constant value of 1 °C, the heat transfer coefficient is 10 W/(m² K). The material of the can can be regarded as ideal heat conductor. The material properties of beer are:

$\rho = 1\,020\,\text{kg/m}^3$, $\lambda = 0.64\,\text{W/(m K)}$ $c_p = 4000\,\text{J/(kg K)}$.

Find

The required cooling time.

Solution

Assumptions

- The material properties of the beer are constant.
- The can is assumed infinitely long.
- The temperature and heat transfer coefficient in the refrigerator are constant.
- The beer in the can does not flow.

Analysis

To determine the *Fourier* number and from there the time, first the dimensionless mean temperature and *Biot* number must be calculated.

$$\overline{\Theta} = \frac{\overline{\vartheta} - \vartheta_\infty}{\vartheta_A - \vartheta_\infty} = \frac{4-1}{30-1} = 0.103 \qquad Bi = \frac{\alpha \cdot R}{\lambda} = \frac{10 \cdot \text{W} \cdot \text{m} \cdot \text{K} \cdot 0.0325 \cdot \text{m}}{\text{m}^2 \cdot \text{K} \cdot 0.64 \cdot \text{W}} = 0.51$$

The *Fourier* number is determined from the bottom diagram in Figure 2.12 as $Fo = 2.5$. To determine the time, the thermal diffusivity must be calculated.

$$a = \frac{\lambda}{\rho \cdot c_p} = \frac{0.64 \cdot \text{W} \cdot \text{m}^3 \cdot \text{kg} \cdot \text{K}}{\text{m} \cdot \text{K} \cdot 1020 \cdot \text{kg} \cdot 4000 \cdot \text{J}} = 1.57 \cdot 10^{-7} \frac{\text{m}^2}{\text{s}}$$

The *Fourier* number according to Equation (2.68) delivers the time.

$$t = \frac{Fo \cdot r^2}{a} = \frac{2.5 \cdot 0.0325^2 \cdot \text{m}^2 \cdot \text{s}}{1.57 \cdot 10^{-7} \cdot \text{m}^2} = 16834 \text{ s} = 4.7 \text{ h}$$

Discussion

This analysis was performed with a lot of assumptions. In reality during the cooling, temperature and heat transfer coefficients change and beer is a liquid and due to the temperature differences and gravity forces a flow will occur in the can. The assumed infinite length of the can is also doubtful. The material of the can itself is really negligible. Despite the assumptions, the calculated time is close to reality. A test I carried out at home resulted in a time of approximately 5 hours. This is a very long time to wait for a cold beer, isn't it!

EXAMPLE 2.12: Cooling of a wire in an oil bath

A wire of 4 mm diameter is drawn from hot steel and cooled in an oil bath of 30 °C temperature. When entering the bath the wire has a temperature of 600 °C. The heat transfer coefficient in the bath is 1 600 W/(m² K). The traveling time of the wire in the bath is 5 s. The material properties of the wire are:

$$\rho = 8000 \text{ kg/m}^3, \lambda = 40 \text{ W/(m K)}, c_p = 800 \text{ J/(kg K)}.$$

Find

The temperatures in the middle and on the surface of the wire, when leaving the bath.

Solution

Schematic   See sketch.

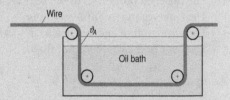

Assumptions

- The material properties of the wire are constant.
- The temperature and heat transfer coefficient in the bath are constant.

Analysis

To find the temperatures, the *Fourier* and *Biot* number have to be calculated. For the *Fourier* number we need the thermal diffusivity.

$$a = \frac{\lambda}{\rho \cdot c_p} = \frac{40 \cdot W \cdot m^3 \cdot kg \cdot K}{m \cdot K \cdot 8000 \cdot kg \cdot 800 \cdot J} = 6.25 \cdot 10^{-6} \frac{m^2}{s}$$

Fourier number: $$Fo = \frac{t \cdot a}{r^2} = \frac{5 \cdot s \cdot 6.25 \cdot 10^{-6} \cdot m^2}{s \cdot 0.002^2 \cdot m^2} = 7.8$$

Biot number: $$Bi = \frac{\alpha \cdot r}{\lambda} = \frac{1600 \cdot W \cdot 0.002 \cdot m \cdot m \cdot K}{m^2 \cdot K \cdot 40 \cdot W} = 0.08$$

The dimensionless surface temperature can be obtained from the top diagram, the temperature in the center from the center diagram in Figure 2.11. We receive: $\Theta_o = 0.26$ and $\Theta_m = 0.27$.

The temperatures are:

$$\vartheta_o = \vartheta_\infty + (\vartheta_A - \vartheta_\infty) \cdot \Theta_o = 30 \text{ °C} + (600 - 30) \cdot K \cdot 0.26 = \mathbf{178.2 \text{ °C}}$$

$$\vartheta_m = \vartheta_\infty + (\vartheta_A - \vartheta_\infty) \cdot \Theta_m = 30 \text{ °C} + (600 - 30) \cdot K \cdot 0.27 = \mathbf{183.9 \text{ °C}}$$

Discussion

In the diagrams we had to use, the reading accuracy is rather poor. Both temperatures obtained have almost the same value. The temperatures can be determined with an accuracy of only ±20 °C.

EXAMPLE 2.13: Boiling eggs

Eggs are boiled by steam condensing on their surface at 100 °C. At the start of boiling, the eggs have a temperature of 20 °C. The heat transfer at condensation was measured with 13 000 W/(m² K). The eggs are assumed to be spheres of homogeneous consistence. The eggs' material properties are:
$\rho = 1050$ kg/m³, $\lambda = 0.5$ W/(m K) $c_p = 3200$ J/(kg K).

Find

a) The temperature in the center reached after 5 minutes of boiling.
b) The time required to reach the same temperature in the center at 2 500 m above sea level, where the condensation temperature drops to 80 °C.

Solution

Schematic   See sketch.

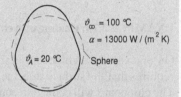

$\vartheta_\infty = 100$ °C
$\alpha = 13000$ W / (m² K)
$\vartheta_A = 20$ °C   Sphere

Assumptions

• The eggs' material properties are constant.
• The eggs are regarded as spheres of homogeneous consistence.
• The temperature and heat transfer coefficient of condensation are constant.

Analysis

a)   To receive the temperature, the *Fourier*, *Biot* number and thermal diffusivity must be determined.

$$a = \frac{\lambda}{\rho \cdot c_p} = \frac{0.5 \cdot W \cdot m^3 \cdot kg \cdot K}{m \cdot K \cdot 1050 \cdot kg \cdot 3200 \cdot J} = 1.49 \cdot 10^{-7} \frac{m^2}{s}$$

Fourier number: $Fo = \dfrac{t \cdot a}{r^2} = \dfrac{300 \cdot s \cdot 1.49 \cdot 10^{-7} \cdot m^2}{s \cdot 0.025^2 \cdot m^2} = 0.071$

Biot number: $Bi = \dfrac{\alpha \cdot r}{\lambda} = \dfrac{13000 \cdot W \cdot 0.025 \cdot m \cdot m \cdot K}{m^2 \cdot K \cdot 0.5 \cdot W} = 650$

The central diagram in Figure 2.13 delivers the dimensionless temperature in the center of the egg with $\Theta_m = 0.75$. The temperature is:

$$\vartheta_m = \vartheta_\infty + (\vartheta_A - \vartheta_\infty) \cdot \Theta_m = 100\ °C + (20 - 100) \cdot K \cdot 0.75 = \mathbf{40\ °C}$$

b)   To determine the time at which the eggs reach a temperature of 40 °C at 80 °C ambient, the dimensionless temperature is needed.

$$\Theta_m = \frac{\vartheta_m - \vartheta_\infty}{\vartheta_A - \vartheta_\infty} = \frac{40 - 80}{20 - 80} = 0.67$$

In the central diagram in Figure 2.13 the *Fourier* number result as $Fo = 0.115$. The boiling time is proportional to the *Fourier* number.

$$t = 0.115 / 0.071 \cdot 5\ \text{min} = \mathbf{8.1\ min}$$

Discussion

Despite the quite simplifying assumptions applied, like sphere shape and homogenous consistence, the results are rather close to reality. Egg white starts to solidify at 42 °C. After 5 minutes of boiling the eggs are still liquid in the center and solidifying to the surface, as a well-boiled egg should be!

EXAMPLE 2.14: Heating up a chipboard

On one side of a chipboard of 20 mm thickness, a veneer is to be glued on. This requires that the surface side, the side the veneer is glued on, is heated up to 150 °C. The chipboard will come into contact with hot air of 200 °C temperature on this side. The other side is placed on an insulator which can be assumed as ideal. At the start of the heating the chipboard has a temperature of 20 °C. The heat transfer coefficient of the air blown with a fan is 50 W/(m² K). The material properties of the chipboard are:

$$\rho = 1\,500\ \text{kg/m}^3,\ \lambda = 1.0\ \text{W/(m K)}\ c_p = 1\,200\ \text{J/(kg K)}.$$

Find

a)   The time to reach the 150 °C surface temperature.
b)   The savings achieved by heating with the insulation on one side.

Solution

Schematic   See sketch.

Chipboard

Insulator

Assumptions

- The material properties of the chipboard are constant.
- On the insulated side of the chipboard no heat is transferred.
- In the air the temperature and heat transfer coefficient are constant.

Analysis

a) On the thermally insulated side of the chipboard no heat transfer occurs. Therefore, the temperature gradient there is zero. The temperature profile is the same as that in one half of a plate with double thickness, heated from both sides. To find the heating time, the *Fourier* number needs to be determined. The dimensionless surface temperature and *Biot* number, calculated with doubled thickness, are required. For the dimensionless surface temperature we receive:

$$\Theta_O = \frac{\vartheta_O - \vartheta_\infty}{\vartheta_A - \vartheta_\infty} = \frac{150 - 200}{20 - 200} = 0.28$$

Biot number:
$$Bi = \frac{\alpha \cdot 2 \cdot s}{\lambda} = \frac{50 \cdot W \cdot 0.02 \cdot m \cdot m \cdot K}{m^2 \cdot K \cdot 1 \cdot W} = 1$$

The *Fourier* number from the top diagram in Figure 2.11 is $Fo = 1.25$. To find the time, first the thermal diffusivity has to be calculated.

$$a = \frac{\lambda}{\rho \cdot c_p} = \frac{1 \cdot W \cdot m^3 \cdot kg \cdot K}{m \cdot K \cdot 1500 \cdot kg \cdot 1200 \cdot J} = 5.56 \cdot 10^{-7} \frac{m^2}{s}$$

The *Fourier* number, Equation (2.68), delivers the time:

$$t = \frac{Fo \cdot (2 \cdot s)^2}{a} = \frac{1.25 \cdot 0.02^2 \cdot m^2 \cdot s}{5.56 \cdot 10^{-7} \cdot m^2} = 900 \text{ s} = 15 \text{ min}$$

b) The heat received from the chipboard per square meter is:

$$\frac{Q}{A} = \frac{m \cdot c_p \cdot (\overline{\vartheta} - \vartheta_A)}{A} = s \cdot \rho \cdot c_p \cdot (\overline{\vartheta} - \vartheta_A)$$

From the bottom diagram in Figure 2.11 we determine the dimensionless mean temperature with $Bi = 1$ and $Fo = 1.25$: $\overline{\Theta} = 0.41$. The mean temperature is:

$$\overline{\vartheta} = \vartheta_\infty + (\vartheta_A - \vartheta_\infty) \cdot \overline{\Theta} = 200 \,°C + (20 - 200) \cdot K \cdot 0.41 = 126.2 \,°C$$

The heat received per square meter we calculate as:

$$Q / A = s \cdot \rho \cdot c_p \cdot (\overline{\vartheta} - \vartheta_A) =$$
$$= 0.02 \cdot m \cdot 1500 \cdot kg/m^3 \cdot 1200 \cdot J/(kg \cdot K) \cdot (126.2 - 20) \cdot K = 3823.2 \text{ kJ} / m^2$$

For the chipboard heated on both sides, the *Fourier* and *Biot* number must be calculated with half of the board thickness. The *Biot* number is then $Bi = 0.5$. The *Fourier* number from the diagram results in $Fo = 2.6$. The heat up time would be shortened to 468 s. The dimensionless mean temperature is: $\overline{\Theta} = 0.31$, i.e. the mean temperature and the heat per square meter: $\overline{\vartheta} = 144.2\,°C$ and $Q/A = \mathbf{4471.2\,kJ/m^2}$.

Discussion

A plane plate insulated on one side can be handled as a plate of double thickness.

2.2.1.3   Special solutions for short periods of time

In Figure 2.10 it can be seen that at time $t = t_1$ the temperature in the middle of the plane plate still has its starting value. In the diagrams 2.11 to 2.13 it can also be seen that with *Fourier* numbers smaller than 0.01 no change of temperature in the middle of the body happens. For a short period of time we receive a special solution of the differential Equation (2.63):

$$\Theta = erf(x^*) + e^{-(x^*)^2} \cdot e^{(x^* + Bi^*)^2} \cdot \left[1 - erf(x^* - Bi^*) \right] \tag{2.71}$$

The *Gauß error function* is *erf* (**er**ror **f**unction), x^* a dimensionless distance to the wall related to $(a \cdot t)^{0.5}$ and Bi^* the *Biot number*, built with the above wall distance x^*.

The dimensionless distance to wall distance x^* and the *Biot* number Bi^* have the following definitions:

$$x^* = \frac{x}{2 \cdot \sqrt{a \cdot t}} \qquad\qquad Bi^* = \frac{\alpha \cdot \sqrt{a \cdot t}}{\lambda} \tag{2.72}$$

Gauß error function:

$$erf(z) = \frac{2}{\sqrt{\pi}} \cdot \int_0^z e^{-x^2} \cdot dx \tag{2.73}$$

The integral can be solved only numerically. The result is shown in Figure 2.14.

For short periods of time the temperature ϑ_m in the middle of the body remains the starting temperature ϑ_A. The surface temperature ϑ_O we receive at $x^* = 0$ is:

$$\Theta_O = e^{Bi^{*2}} \cdot \left[1 - erf(Bi^*) \right] \tag{2.74}$$

$$\Theta_O = \frac{1}{\sqrt{\pi} \cdot Bi^*} \tag{2.75}$$

The errors when using the error function solution are less than 1 % if compared to an exact solution for short periods of time.

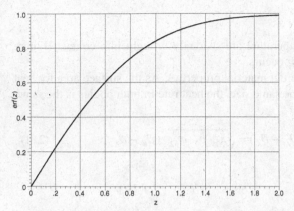

Figure 2.14: *Gauß* error function

The surface temperatures for Bi^* reaching infinite values, is the limit of Equation (2.71):

$$\Theta_O = erf(x^*) = erf\left(\frac{x}{2\cdot\sqrt{a\cdot t}}\right) \tag{2.76}$$

The heat flux through the surface is:

$$\dot{q}_O(t) = \frac{\lambda}{\sqrt{\pi\cdot a\cdot t}}\cdot(\vartheta_A - \vartheta_O) = \frac{\sqrt{\lambda\cdot\rho\cdot c_p}}{\sqrt{\pi\cdot t}}\cdot(\vartheta_A - \vartheta_O) \tag{2.77}$$

The heat transferred through the surface until time t can be determined by integrating Equation (2.77) from zero to t.

$$Q_O(t) = A\cdot\int_0^t \dot{q}_O(t)\cdot dt = \frac{2\cdot A\cdot\lambda\cdot t}{\sqrt{\pi\cdot a\cdot t}}\cdot(\vartheta_A - \vartheta_O) \tag{2.78}$$

With an energy balance equation the heat transferred in the time t can be determined with Equation (2.70).

$$Q = V\cdot\rho\cdot c_p\cdot(\vartheta_A - \overline{\vartheta}) \tag{2.79}$$

With the combination of the above equations the mean temperature results in:

$$(\vartheta_A - \overline{\vartheta}) = \frac{2\cdot A\cdot\lambda\cdot t}{V\cdot\rho\cdot c_p\cdot\sqrt{\pi\cdot a\cdot t}}\cdot(\vartheta_A - \vartheta_O) = \frac{2\cdot A\cdot\sqrt{a\cdot t}}{V\cdot\sqrt{\pi}}\cdot(\vartheta_A - \vartheta_O) \tag{2.80}$$

The surface area to volume ratio of a plane plate is $1/s$ (considered that s is the half of the plate thickness), of the cylinder $4/d$ and of the sphere $6/d$. With these values and the surface temperature determined by Equation (2.76), the mean temperature can be determined.

Equations (2.76) to (2.80) are valid only if the *Biot* number has large values.

2.2.2   Coupled systems

If two bodies of different temperature are suddenly brought into contact (Figure 2.15), a *contact temperature* ϑ_K occurs.

Both bodies may have different material properties. As the surface area for heat transfer for both bodies is of the same size, the heat rate in both bodies is the same. Equation (2.77) delivers:

$$\sqrt{\lambda_1 \cdot \rho_1 \cdot c_{p1}} \cdot (\vartheta_{A1} - \vartheta_K) = \sqrt{\lambda_2 \cdot \rho_2 \cdot c_{p2}} \cdot (\vartheta_K - \vartheta_{A2}) \qquad (2.81)$$

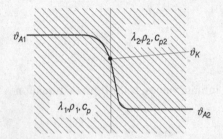

Figure 2.15: Contact of two bodies with different temperatures

Transforming this equation yields the contact temperature ϑ_K :

$$\vartheta_K = \left(\vartheta_{A1} + \sqrt{\frac{\lambda_2 \cdot \rho_2 \cdot c_{p2}}{\lambda_1 \cdot \rho_1 \cdot c_{p1}}} \cdot \vartheta_{A2} \right) \cdot \left(1 + \sqrt{\frac{\lambda_2 \cdot \rho_2 \cdot c_{p2}}{\lambda_1 \cdot \rho_1 \cdot c_{p1}}} \right)^{-1} \qquad (2.82)$$

With Equation (2.82), it can be demonstrated why we feel different bodies of the same temperature differently "warm" or "cold". The contact temperatures are governed by the ratio of the heat *penetration coefficient* $(\lambda \cdot \rho \cdot c_p)^{0.5}$ of both bodies. A copper plate has a much higher heat penetration coefficient than the human hand. Therefore we feel almost the full temperature of the copper plate. With a styrofoam plate it is the contrary: we feel the temperature of our hand.

As long as the temperature change does not reach deeper areas of the body, the contact temperature remains constant. If deeper areas are reached, a temperature homogenization starts. This can be illustrated by touching a very thin hot aluminum sheet: In the first moment, the contact temperature develops, this is the temperature of the aluminum sheet. However, as the sheet is very thin and has a very small mass, the temperature homogenization starts rapidly and the contact temperature does not prevail. The small thermal energy stored in the small mass of the thin aluminum sheet is not able to increase the temperature of the skin as deep as the nerves would feel the hot temperature.

Touching a thicker aluminum plate of the same temperature, we would feel the hot temperature and burn our fingers.

EXAMPLE 2.15: Felt temperatures

A thick styrofoam and copper plate have the same temperature of 0 °C. They are touched by a human hand. Material properties:

Styrofoam:	$\rho = 15\,\text{kg/m}^3$	$\lambda = 0.029\,\text{W/(m K)}$	$c_p =$	1250 J/(kg K)
Copper:	$\rho = 8\,300\,\text{kg/m}^3$	$\lambda = 372\,\text{W/(m K)}$	$c_p =$	419 J/(kg K)
Hand:	$\rho = 1\,020\,\text{kg/m}^3$	$\lambda = 0.5\,\text{W/(m K)}$	$c_p =$	2400 J/(kg K)

Find

The contact temperatures.

Solution

Analysis

The contact temperatures will be determined with Equation (2.82). First the heat penetration coefficient will be calculated for all materials and named ξ.

Styrofoam: $\quad \xi_{styrofoam} = \sqrt{\rho \cdot c_p \cdot \lambda} = \sqrt{15 \cdot 1250 \cdot 0.029} = 23.3$

Copper: $\quad \xi_{copper} = \sqrt{\rho \cdot c_p \cdot \lambda} = \sqrt{8300 \cdot 419 \cdot 372} = 35968$

Hand: $\quad \xi_{hand} = \sqrt{\rho \cdot c_p \cdot \lambda} = \sqrt{1020 \cdot 2400 \cdot 0.5} = 1106$

The contact temperature between styrofoam and hand is:

$$\vartheta_{K,\,styrofoam-hand} = \frac{\vartheta_{A,\,hand} + (\xi_{styrofoam} / \xi_{hand}) \cdot \vartheta_{A,\,styrofoam}}{1 + (\xi_{styrofoam} / \xi_{hand})} =$$

$$= \frac{36\,°C + (23.3/1106) \cdot 0\,°C}{1 + 23.3/1106} = \mathbf{35.26\,°C}$$

The contact temperature between copper and hand is:

$$\vartheta_{K,\,copper-hand} = \frac{\vartheta_{A,\,hand} + (\xi_{copper} / \xi_{hand}) \cdot \vartheta_{A,\,copper}}{1 + \xi_{copper} / \xi_{hand}} =$$

$$= \frac{36\,°C + (35968/1106) \cdot 0\,°C}{1 + 35968/1106} = \mathbf{1.07\,°C}$$

Discussion

Because of the very low thermal penetration coefficient of styrofoam, the temperature of 0 °C is felt as hand-hot, the copper is felt as ice-cold.

2.2.3   Special cases at $Bi = 0$ and $Bi = \infty$

At a very small *Biot* number (near zero), i.e. the outer heat transfer coefficient is very small or the body has a very large thermal conductivity, the temperature in the body is independent of the location: it is only a function of time.

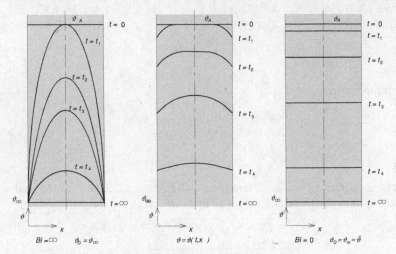

Figure 2.16: Temperature distribution in special cases

At very large *Biot* numbers, i.e. very large outside heat transfer coefficients or very low thermal conductivity of the body, the surface temperature ϑ_0 has the value of the temperature of the surroundings ϑ_∞.

This is illustrated in Figure 2.16. The left figure shows the temperature distribution for $Bi = \infty$. The surface temperature is equal to the temperature of the surroundings. The right figure is for $Bi = 0$, the temperature in the body changes only with time. The figure in the center represents the temperature distribution at finite *Biot* numbers.

2.2.4   Temperature changes at small *Biot* numbers

In many technical processes not the temperature distribution but the change of the mean temperature during cooling or heating is of interest. As the diagrams in Figures 2.11 to 2.13 show, with small *Biot* number – i.e. the outside heat transfer coefficient is much lower than that inside the body – the temperature difference between the mean temperature and that in the center of the body is rather small. For *Biot* numbers smaller than 0.5, the mean temperature of the body may be assumed as the *spacial uniform temperature (lumped capacitance method)*.

For *Biot* numbers smaller than 1, the lumped capacitance method is possible, but is a rather rough approximation.

2.2.4.1 A small body immersed in a fluid with large mass

A small body with the mass of m_1, specific heat capacity c_{p1} and temperature ϑ_{A1} is immersed in a fluid with the temperature ϑ_{A2} (Figure 2.17).

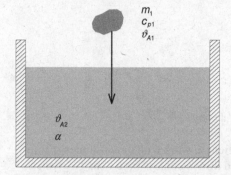

Figure 2.17: A small body immersed in a fluid of large mass

The fluid mass and therefore its heat capacity, can be assumed as large, if the temperature change of the fluid is negligible when immersing the small body . The heat transfer coefficient on the fluid side of the surface of the body is α. The wall of the container of the fluid is insulated. The heat rate transferred from the body to the fluid is the transient change of the enthalpy of the body.

$$\dot{Q} = -\frac{dH}{dt} = -m_1 \cdot c_{p1} \cdot \frac{d\vartheta_1}{dt} \tag{2.83}$$

The heat rate can also be determined with the rate equation.

$$\dot{Q} = \alpha \cdot A \cdot (\vartheta_1 - \vartheta_{A2}) \tag{2.84}$$

Both equations deliver the differential equation.

$$\frac{d\vartheta_1}{dt} = -\frac{\alpha \cdot A}{m_1 \cdot c_{p1}} \cdot (\vartheta_1 - \vartheta_{A2}) \tag{2.85}$$

With the temperature ϑ_1 substituted by $\vartheta_1 - \vartheta_{A2}$, we receive after separation of the variables:

$$\frac{d(\vartheta_1 - \vartheta_{A2})}{\vartheta_1 - \vartheta_{A2}} = -\frac{\alpha \cdot A}{m_1 \cdot c_{p1}} \cdot dt \tag{2.86}$$

Assumed constant heat transfer coefficients and material properties, the equation can be integrated.

$$(\vartheta_1 - \vartheta_{A2}) = (\vartheta_{A1} - \vartheta_{A2}) \cdot e^{-\frac{\alpha \cdot A}{m_1 \cdot c_{p1}} \cdot t} \tag{2.87}$$

Figure 2.18 shows the transient temperature change of the body, which is reaching asymptotically the fluid temperature.

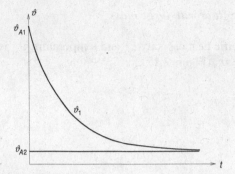

Figure 2.18: Temperature progression of a cooled body

For a generalized description, it is convenient to work with dimensionless parameters. The dimensionless temperature Θ and dimensionless time τ can be given as:

$$\Theta = \frac{\vartheta_{A1} - \vartheta_1}{\vartheta_{A1} - \vartheta_{A2}} = 1 - e^{-\tau} \qquad \text{with:} \qquad t_0 = \frac{m_1 \cdot c_{p1}}{\alpha \cdot A} \qquad \tau = \frac{t}{t_0}$$

The time t_0 is the time in which the immersed body changes its temperature by 1 K with a heat rate produced by the heat transfer at 1 K temperature difference. Figure 2.19 shows the dimensionless representation of the temperature progression.

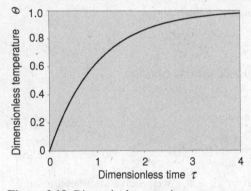

Figure 2.19: Dimensionless transient temperature progress

EXAMPLE 2.16: Quenching of steel parts

Cylindrical steel parts with a mass of 1.2 kg and surface area of 300 cm² should be cooled from 800 °C to 300 °C in an oil bath of 50 °C temperature. The heat transfer coefficient is 600 W/(m² K).

Material properties of steel: $\lambda = 47$ W/(m K), $c_p = 550$ J/(kg K).

Find

The time required to reach the temperature of 300 °C.

Solution

Assumptions

- The material properties are constant.
- The temperature of the steel part has no spacial differences.
- The temperature and heat transfer coefficients of the oil bath are constant.

Analysis

First the value of the *Biot* number is determined.

$$Bi = \alpha \cdot r / \lambda = 600 \cdot 0.01 / 47 = 0.128$$

It is smaller than 0.5. Therefore Equation (2.87) can be applied to determine the time.

$$t = \frac{m_1 \cdot c_{p1}}{\alpha \cdot A} \ln\left(\frac{\vartheta_{A1} - \vartheta_{A2}}{\vartheta_1 - \vartheta_{A2}}\right) = \frac{1.2 \cdot kg \cdot 550 \cdot J \cdot m^2 \cdot K}{kg \cdot K \cdot 600 \cdot W \cdot 0.03 \cdot m^2} \cdot \ln\left(\frac{800 - 50}{300 - 50}\right) = \mathbf{40.2 \ s}$$

Discussion

This calculation is rather simple. The exact value determined with the diagram in Figure 2.12, is 46.8 s. The error is 14 %. It has to be checked if this error is acceptable in view of the requirements of the quenching process.

2.2.4.2   *A body is immersed into a fluid of similar mass*

Figure 2.20 shows the dipping of a body into a fluid in an insulated basin. The fluid mass is similar to the mass of the body. The fluid mass is m_2 and its specific heat capacity c_{p2}. The mass of the body is m_1 and its specific heat capacity c_{p2}. At the moment of immersing, the body's temperature is ϑ_{A1}, that of the fluid ϑ_{A2}.

Depending on the magnitude of the product of mass and heat capacity $m \cdot c_p$ of the body and the fluid, a change of the fluid temperature will occur.

The heat rate from the body is the transient change of its enthalpy.

$$-\dot{Q} = m_1 \cdot c_{p1} \cdot \frac{d\vartheta_1}{dt} \tag{2.88}$$

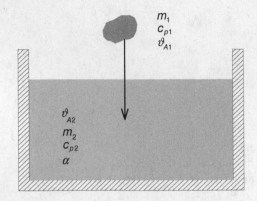

Figure 2.20: A body is dipped into a fluid of comparable mass

The heat rate equals the transient change of the enthalpy of the fluid:

$$\dot{Q} = m_2 \cdot c_{p2} \cdot \frac{d\vartheta_2}{dt} \qquad (2.89)$$

The heat rate is also defined by the rate equation.

$$\dot{Q} = \alpha \cdot A \cdot (\vartheta_1 - \vartheta_2) \qquad (2.90)$$

With Equations (2.88) and (2.89) the change of the temperature difference between body and fluid can be determined.

$$d(\vartheta_1 - \vartheta_2) = -\dot{Q} \cdot \left[\frac{1}{m_1 \cdot c_{p1}} + \frac{1}{m_2 \cdot c_{p2}} \right] \cdot dt \qquad (2.91)$$

From Equation (2.90) the heat rate can be inserted into Equation (2.91).

$$\frac{d(\vartheta_1 - \vartheta_2)}{(\vartheta_1 - \vartheta_2)} = \alpha \cdot A \cdot \left[\frac{1}{m_1 \cdot c_{p1}} + \frac{1}{m_2 \cdot c_{p2}} \right] \cdot dt \qquad (2.92)$$

With the assumption that the masses, heat transfer coefficients, specific heat capacities are constant, the differential equation can be solved.

$$\ln \frac{(\vartheta_1 - \vartheta_2)}{(\vartheta_{A1} - \vartheta_{A2})} = -\alpha \cdot A \cdot \left[\frac{1}{m_1 \cdot c_{p1}} + \frac{1}{m_2 \cdot c_{p2}} \right] \cdot t \qquad (2.93)$$

The temperature difference between body and fluid is:

$$\vartheta_1 - \vartheta_2 = (\vartheta_{A1} - \vartheta_{A2}) \cdot \exp\left[-\alpha \cdot A\left(\cdot \frac{1}{m_1 \cdot c_{p1}} + \frac{1}{m_2 \cdot c_{p2}} \right) \cdot t \right] \qquad (2.94)$$

With Equation (2.94), the temperature difference between body and fluid can be determined, but neither that of the body nor that of the fluid. The temperature of the body and of the fluid can be calculated with Equations (2.88) and (2.89). Assuming constant masses and specific heat capacities, the heat transferred in the time t results as:

$$Q(t) = m_1 \cdot c_{p1} \cdot (\vartheta_{A1} - \vartheta_1)$$
$$Q(t) = m_2 \cdot c_{p2} \cdot (\vartheta_2 - \vartheta_{A2}) \tag{2.95}$$

Equation (2.95) delivers the temperature reached in infinite time.

$$\vartheta_\infty = \frac{m_1 \cdot c_{p1} \cdot \vartheta_{A1} + m_2 \cdot c_{p2} \cdot \vartheta_{A2}}{m_1 \cdot c_{p1} + m_2 \cdot c_{p2}} \tag{2.96}$$

The combination of Equations (2.95) and (2.96) delivers for the temperatures:

$$\frac{\vartheta_{A1} - \vartheta_1}{\vartheta_2 - \vartheta_{A2}} = \frac{\vartheta_{A1} - \vartheta_\infty}{\vartheta_\infty - \vartheta_{A2}} = \frac{m_2 \cdot c_{p2}}{m_1 \cdot c_{p1}} \tag{2.97}$$

Equation (2.97) can be solved for ϑ_1 or ϑ_2 and inserted in Equation (2.94). For the temperature of the body and of the fluid, we receive:

$$\vartheta_1 - \vartheta_\infty = (\vartheta_{A1} - \vartheta_\infty) \cdot \exp\left[-\alpha \cdot A \cdot \left(\frac{1}{m_1 \cdot c_{p1}} + \frac{1}{m_2 \cdot c_{p2}}\right) \cdot t\right] \tag{2.98}$$

$$\vartheta_2 - \vartheta_\infty = (\vartheta_{A2} - \vartheta_\infty) \cdot \exp\left[-\alpha \cdot A \cdot \left(\frac{1}{m_1 \cdot c_{p1}} + \frac{1}{m_2 \cdot c_{p2}}\right) \cdot t\right] \tag{2.99}$$

The temporal temperature distribution is illustrated in Figure 2.21.

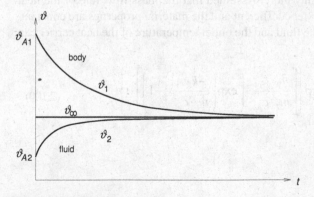

Figure 2.21: Temperature progress in the body and fluid

This type of thermal conduction calculation is not very realistic, as it assumes that there are no spacial temperature differences in fluid and body. This could only be realized when both the body and the fluid would have infinite thermal conductivity. However, for a few cases the calculation procedure described can deliver rough estimates.

2.2.4.3   Heat transfer to a static fluid by a flowing heat carrier

Figure 2.22 shows a tank insulated to the environment, containing a fluid which is heated or cooled by a flowing heat carrier.

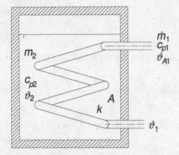

Figure 2.22: Heating or cooling a fluid by means of a flowing heat carrier

Between the fluid and the heat carrier there is a thermally conducting wall, assumed to have infinite thermal conductivity. The temperature of the fluid at the time $t = 0$ is ϑ_{A2}, that of the heat carrier is ϑ_{A1}. The temperature of the heat carrier, when leaving the fluid, is ϑ_1. The heat carrier is heated or cooled by the fluid. The heat transfer occurs with a constant heat transfer coefficient. After infinite time the fluid will reach the inlet temperature of the heat carrier. From then on, the temperature of the carrier will not change any more. Assumed that the mass flow rate of the heat carrier, the overall heat transfer coefficient and the material properties are constant, the temperature change of the fluid and the outlet temperature of the heat carrier can be determined.

$$\vartheta_{A1} - \vartheta_2 = (\vartheta_{A1} - \vartheta_{A2}) \cdot \exp\left\{\left[\frac{\dot{m}_1 \cdot c_{p1}}{m_2 \cdot c_{p2}} \cdot \left(\exp\left(\frac{-k \cdot A}{\dot{m}_1 \cdot c_{p1}}\right) - 1\right)\right] \cdot t\right\} \qquad (2.100)$$

EXAMPLE 2.17: Cooling of a wire in a water bath

A wire with 2 mm diameter and a temperature of 300 °C is cooled while traveling through a water bath. The wire velocity is 0.5 m/s and its traveling length 5 meters. The water has a mass of 5 kg and at the beginning of the process a temperature of 20 °C. Material properties of the wire: $\rho = 8\,000\ \text{kg/m}^3$, $\lambda = 47\ \text{W/(m K)}$, $c_p = 550\ \text{J/(kg K)}$. The heat transfer coefficient in the water is 1 200 W/(m² K) and the specific heat capacity $c_{p2} = 4\,192$ (J/kg K).

Find

After which time the water must be changed, to avoid a wire temperature of more than 100 °C.

Solution

Assumption

- The material properties are constant.
- The inlet temperature of the wire is constant.
- The heat transfer coefficient oft the bath is constant.
- There are no spacial temperature differences in the water bath.

Analysis

With Equation (2.87) we can determine at what water temperature the wire temperature exceeds 100 °C. The wire is the flowing heat carrier. The temperature of the bath at 100 °C wire temperature we calculate with Equation (2.87). The traveling time of the wire in the water is 10 s. In Equation (2.87) it can be seen, that the length of the wire has no influence with regard to cooling. In the exponent in the term A/m_1 the length can be eliminated. This is correct as long as the cross-section at the ends are irrelevant.

$$(\vartheta_1 - \vartheta_{A2}) = (\vartheta_{A1} - \vartheta_{A2}) \cdot e^{-\frac{\alpha \cdot A}{m_1 \cdot c_{p1}} \cdot t} = (\vartheta_{A1} - \vartheta_{A2}) \cdot \exp\left(-\frac{4 \cdot \alpha \cdot \pi \cdot d \cdot l}{\rho \cdot \pi \cdot d^2 \cdot l \cdot c_{p1}} \cdot t\right) =$$

$$= (\vartheta_{A1} - \vartheta_{A2}) \cdot \exp\left(-\frac{4 \cdot \alpha}{\rho \cdot d \cdot c_{p1}} \cdot t\right)$$

With a given outlet temperature $\vartheta_1 = 100$ °C the temperature ϑ_{A2} can be determined.

$$\vartheta_{A2} = \frac{\vartheta_{A1} \cdot \exp\left(-\dfrac{4 \cdot \alpha}{\rho \cdot d \cdot c_{p1}} \cdot t\right) - \vartheta_1}{\exp\left(-\dfrac{4 \cdot \alpha}{\rho \cdot d \cdot c_{p1}} \cdot t\right) - 1} = \frac{300 \cdot e^{-\frac{4 \cdot 1200}{8000 \cdot 0.002 \cdot 550} \cdot 10} - 100}{e^{-\frac{4 \cdot 1200}{8000 \cdot 0.002 \cdot 550} \cdot 10} - 1} \cdot {}^\circ\text{C} = 99.14\ {}^\circ\text{C}$$

With Equation (2.100) the time required to heat the water from 20 °C to 99,14 °C can be calculated. First the mass flow rate and the surface area has to be calculated.

$$\dot{m}_1 = c \cdot 0.25 \cdot \pi \cdot d^2 \cdot \rho_1 =$$

$$= 0.5 \cdot \text{m/s} \cdot 0.25 \cdot \pi \cdot 0.002^2 \cdot \text{m}^2 \cdot 8000 \cdot \text{kg/m}^3 = 0.01257\ \text{kg/s}$$

$$A = \pi \cdot d \cdot l = \pi \cdot 0.002 \cdot \text{m} \cdot 5 \cdot \text{m} = 0.0314\ \text{m}^2$$

$$t = \frac{\ln \dfrac{\vartheta_{A1} - \vartheta_2}{\vartheta_{A1} - \vartheta_{A2}}}{\dfrac{\dot{m}_1 \cdot c_{p1}}{m_2 \cdot c_{p2}} \cdot \left[e^{\left(\frac{-\alpha \cdot A}{\dot{m}_1 \cdot c_{p1}}\right)} - 1 \right]} = \frac{\ln \dfrac{300 - 99.14}{300 - 20}}{\dfrac{0.01257 \cdot 550}{5 \cdot 4192} \cdot \left[e^{\left(\frac{-1200 \cdot 0.0314}{0.01257 \cdot 550}\right)} - 1 \right]} = \mathbf{1012\ s}$$

Discussion

This example shows that even solid bodies can have a mass flow rate when they are moving. This calculation again assumes that in the water there is only a temporal but no spacial temperature change. With the rather long time the result is close to reality, because the traveling wire mixes the water and therefore minimizes the temperatures. However, this is certainly not a solely conduction process.

2.2.5   Numerical solution of transient thermal conduction equations

In Chapter 2.2.1.1 the differential equation for transient thermal conduction (2.64) was developed. A series of solutions for simple geometries, for which a one-dimensional thermal conduction can be assumed, was presented. To determine the temperature distribution in complicated geometries with multidimensional thermal conduction , numerical methods, implemented in computer codes, are used. Some of the basic fundamentals of numeric solution, based on the finite difference method, are discussed here.

2.2.5.1   *Discretization*

For the treatment we use the example of a cylindric rod with the start temperature of ϑ_A, adiabatic or ideally thermally insulated on the lateral surface and on one end. At the time $t_0 = 0$ it is brought in contact with another body of the constant temperature ϑ_∞. Figure 2.23 illustrates the temperature distribution at points in time *with t ≥ 0*.

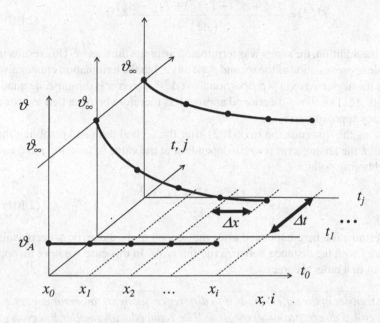

Figure 2.23: Temporal development of the temperature distribution in an adiabatic rod

Instead of using the continuous temperature distribution $\vartheta(x,t)$ the discrete temperatures $\vartheta(x_i, t_j)$ at a limited number of points in time and space can be determined. Such points and the corresponding temperature values are also plotted in Figure 2.23. The mathematical procedures needed for the determination of the temperatures in these discrete points are substantially simpler than the analytical solution of the differential equation of the transient thermal conduction, as will be shown hereafter.

For this first the differential equation must be transferred into a discretized form. The mathematical base for this transfer is a *Taylor* series for a function f:

$$f(\xi + \Delta\xi) = \sum_{n=0}^{\infty} \frac{\partial^n f}{\partial \xi^n}\bigg|_\xi \frac{(\Delta\xi)^n}{n!} \tag{2.101}$$

Writing this series up to the fourth term for the left and right neighbor of ξ, we receive:

$$f(\xi + \Delta\xi) = f(\xi) + f'(\xi)\Delta\xi + f''\frac{(\Delta\xi)^2}{2} + f'''\frac{(\Delta\xi)^3}{6} + ...$$

$$f(\xi - \Delta\xi) = f(\xi) - f'(\xi)\Delta\xi + f''\frac{(\Delta\xi)^2}{2} - f'''\frac{(\Delta\xi)^3}{6} + \tag{2.102}$$

The addition of these two equations provides a formula for the second derivative, which depends only on the values of the function at ξ and its direct neighbors:

$$f''(\xi) = \frac{f(\xi + \Delta\xi) - 2f(\xi) + f(\xi - \Delta\xi)}{(\Delta\xi)^2} \qquad (2.103)$$

Besides the addition, the series was terminated after the third term. This results in a very simple approximation of the second derivative, but the termination creates also an error. As the neglected term is proportional to $(\Delta\xi)^2$, the error is diminishing quadratically with $\Delta\xi$. The above-mentioned approach is therefore being called a second order accuracy approximation.

Terminating the first equation in (2.102) after the second term and resolving the derivative of f, the arising term is again dependent of the value of function f in ξ and in one neighboring point:

$$f'(\xi) = \frac{f(\xi + \Delta\xi) - f(\xi)}{\Delta\xi} \qquad (2.104)$$

As the termination here happened after the second term, the error is decreasing proportionally with the distance between the intervals. In this case we have an approximation of first order accuracy.

> *With the transfer of the differentials into difference quotients, an error occurs, which is called discretization error. It will be reduced with smaller intervals between the discrete points.*

As in Equation (2.104) only the value of f in the neighboring point in positive ξ-direction and the value of f at ξ are present, the expression is called downwind difference quotient. An upwind difference quotient can be given by transforming the first version of equation (2.102) in a similar manner. The term in Equation (2.103) is called central difference quotient, as here the value of the function at ξ as well as in both neighboring points are included.

By now, the differential equation of transient thermal conduction as defined in Equation (2.63)

$$\frac{\partial\vartheta}{\partial t} = a \cdot \frac{\partial^2\vartheta}{\partial x^2}$$

can be transformed into a set of discrete difference equations which is valid for the rod discussed here by inserting the approximations according to Equations (2.103) and (2.104):

$$\left.\frac{\vartheta(t + \Delta t) - \vartheta(t)}{\Delta t}\right|_x = a \cdot \left.\frac{\vartheta(x + \Delta x) - 2\vartheta(x) + \vartheta(x - \Delta x)}{(\Delta x)^2}\right|_t \qquad (2.105)$$

To achieve this difference equation, in Equations (2.103) and (2.104) the general coordinate ξ was replaced by the spacial coordinate x and the time t. Furthermore, equal distances between all spacial points and also between all points in time are assumed, which results in a so-called equidistant mesh. The index x given on the left side of the equation and t on the right side, expresses that the temporal differencing is done at point x and the spacial differencing is performed at time t.

2.2.5.2   *Numerical solution*

To be able to efficiently write and solve difference equation sets for a large number of discrete points i, j, an index notation is implemented. The equidistant spacial points have the index i, the equidistant points in time the index j. The index notification for the spatial distribution has the following appearance:

$$\vartheta(x_i + \Delta x) = \vartheta(x_{i+1}) = \vartheta_{i+1} \qquad \vartheta''(x_i) = \vartheta'' \qquad (2.106)$$

For the time domain only x is replaced by t and i by j. Equation (2.105) in index notification has the following appearance:

$$\frac{\vartheta_{i,j+1} - \vartheta_{i,j}}{\Delta t} = a \cdot \frac{\vartheta_{i+1,j} - 2\vartheta_{i,j} + \vartheta_{i-1,j}}{(\Delta x)^2} \qquad (2.107)$$

This equation can explicitly be solved for $\vartheta_{i,j+1}$:

$$\vartheta_{i,j+1} = a \cdot \Delta t \frac{\vartheta_{i+1,j} - 2\vartheta_{i,j} + \vartheta_{i-1,j}}{(\Delta x)^2} + \vartheta_{i,j} \qquad (2.108)$$

A closer look reveals that the right hand side of this equation contains spacial temperature values at time j only. If the spacial distribution of the temperature in time plane j is known, the temperatures in the plane $j + 1$ can be calculated by the given equation in the most simple manner. From a known starting distribution of the temperature at $j=0$ with successively "moving forward" in time, the spacial temperature distribution from one time plane to the next can be calculated. As the difference quotient used in space is a central one, the process is well known as FTCS-procedure (Forward Time Center Space).

Figure 2.24 shows a projection of the graph in Figure 2.23 on the t-x-plane, i.e. a top-view of the diagram. The procedure for the determination of a temperature in the next time plane is illustrated here. The presentation also reveals why the expression "calculation grid" is used, when discussing numerical solution procedures of partial differential equations.

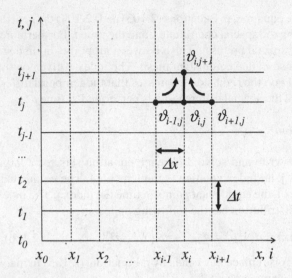

Figure 2.24: Graphical representation of the FTCS-procedure

A partial differential equation can be transformed into a difference equation, which can be solved with simple mathematical methods. The transformation procedure is called discretization. The difference equation must be solved in a large number of points in time and space, which is usually done by using computer codes.

The transfer of this method to multidimensional areas and geometries is relatively simple. For this, only the spacial derivative in the additional directions y and z must be added on the right hand side of Equation (2.63). With this, terms with the index k and l have to be added in Equations (2.107) and (2.108). The procedure for the determination of the spacial temperature distribution in the next time plane remains the same.

In contrast to analytical methods solving the differential equations of transient heat transfer, the numerical methods can be applied to complicated multidimensional geometries too.

2.2.5.3   Selection of the grid spacing and of the time interval

Above it was shown that with the discretization an approximation error is caused in the transient thermal conduction difference equation and in its numeric solution. The This error can be minimized by reducing the grid spacing and the interval between discrete points in time. However, there is an important relationship between grid width and time interval, called *stability criterion*.

$$\Delta t \leq \frac{(\Delta x)^2}{2a}$$

(2.109)

If the time interval exceeds this value, the procedure will not provide useful solutions, but will diverge. Therefore, the selection of a smaller spacing requires a smaller time interval. This relationship is rigorously valid for the explicit Finite Difference Method presented here.

There are methods, however, that reduce the requirements on temporal spacing. Consider – as an example – the application of an upwind difference for the time discretization. This will result in a implicit difference equation, i.e. the determination of the spacial temperature distribution in the next time interval requires the solution of an algebraic equation system.

This is possible mathematically, but more complicated than the method presented here. Anyhow the implicit method is, because of it better stability and the resulting larger time interval, implemented in commercial numerical computer codes. With this, an efficient numerical solution is possible for large areas with a huge number of grid nodes.

There exists a large number of other methods for the spacial and temporal discretization and for solving the created equation systems. Examples are the Finite Elements Method (FEM) and the Finite Volume Method (FVM). With these methods the disadvantages of the Finite-Difference-Method can be circumvented.

3 Forced convection

Forced convection is ruled by the temperature distribution and by a fluid motion, caused by an outer pressure difference. The latter can be established by a pump or by a fluid column. Forced convection is the type of heat transfer most commonly occur in industrial applications. In the heat exchanger the heat is transferred between two fluids separated by a wall. Our task in this chapter will be to determine the heat transfer coefficients of heat exchangers as a function of flow conditions, geometry and temperature differences.

Let us consider a fluid of the temperature ϑ_F, that flows in a pipe with a wall temperature ϑ_W, the heat flux at any location is as defined in Chapter 1.1.1.2:

$$\dot{q} = \alpha \cdot (\vartheta_F - \vartheta_W) \tag{3.1}$$

Within this definition, the temperature of the fluid is a constant value over the cross-section of the pipe but it can vary along the pipe. Experience shows (and Fourier's law for heat transfer requires) that a temperature profile develops close to the wall. In turbulent flows close to the wall a *thermal boundary layer* exists [3.1], in which the temperature varies from wall to fluid temperature (Figure 3.1).

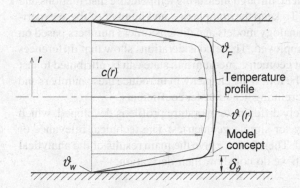

Figure 3.1: Temperature profile in a turbulent pipe flow

The heat is transferred through the boundary layer by conduction. Keeping this in mind and extending Equation (3.1) through building a balance at the wall surface, the following equation results:

$$\dot{q} = \alpha \cdot \left(\vartheta_F - \vartheta_W \right) = -\lambda \left(\frac{\partial \vartheta}{\partial r} \right)_{r_W}$$

$$\alpha = -\lambda \frac{\left(\dfrac{\partial \vartheta}{\partial r} \right)_{r_W}}{\vartheta_F - \vartheta_W} \tag{3.2}$$

Equation (3.2) shows that the heat transfer coefficient depends on the temperature distribution in the flowing medium, and the heat conductivity. The temperature distribution in the fluid is connected in a complex way with the velocity distribution. However, a linearization of the temperature distribution close to the wall surface leads to an approximation for the relationship between temperature drop in the boundary layer, the heat transfer coefficient, the thermal conductivity and the boundary layer thickness δ_ϑ:

$$\left(\frac{\partial \vartheta}{\partial r} \right)_{r_W} \approx \frac{\vartheta_F - \vartheta_W}{\delta_\vartheta}$$

$$\alpha \approx \frac{\lambda}{\delta_\vartheta} \quad \text{and} \quad \delta_\vartheta = \frac{\lambda}{\alpha} \tag{3.3}$$

Because of its small dimension, the thickness of the thermal boundary layer cannot be measured directly in many technical applications. The boundary layer would be disturbed by the measurement. Unfortunately, the boundary layer thickness cannot be determined analytically in most cases. In all these cases, an indirect determination of the heat transfer coefficient, through measuring temperature distributions and heat rates, is the only solution. To keep the complexity and number of these experiments in an acceptable range, analogy models and dimensionless numbers based on similarity considerations are employed. These considerations show that differences in the heat transfer for different geometry, medium and state can be attributed to just a few characteristic numbers. The following chapter will introduce these numbers and explain their application in determining heat transfer coefficients.

In laminar flows, a completely different temperature profile is developed, which can be determined analytically for simple geometries. The technical relevance of these cases, however, is limited. Therefore we give the main results of the analytical solutions in this book, although we do not show their derivation.

3.1   Dimensionless parameters

The transfer processes governing convective heat transfer can be described using partial differential equations. Basically the velocity (three components), pressure and temperature must be determined. For their determination we need five equations, which we can derive from the conservation or balance equations:

- Conservation of mass – mass balance – continuity equation
- Conservation of momentum – momentum balance – equation of motion
- Conservation of energy – energy balance – energy equation

In the following paragraphs, we demonstrate in a simplified way how the important characteristic dimensionless numbers are connected with these equations and in which way they can be used to determine the heat transfer coefficients.

3.1.1   Continuity equation

First we consider a fixed cuboid with the edge lengths dx, dy, dz in a Cartesian coordinate system (see Figure 3.2).

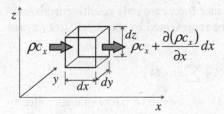

Figure 3.2: Derivation of the continuity equation

The mass flowing into the cuboid through the cross-section $dy \cdot dz$ is

$$\dot{m}_x = \rho \cdot c_x \cdot dy \cdot dz$$

and the mass leaving the cuboid through the adjacent cross-section is given by:

$$\dot{m}_{x+dx} = \rho \cdot c_x \cdot dy \cdot dz + \frac{\partial (\rho \cdot c_x)}{\partial x} dx \cdot dy \cdot dz$$

For an incompressible fluid with the density ρ, the difference of both mass flow rates will be:

$$d\dot{m}_x = -\frac{\partial c_x}{\partial x} \rho \cdot dx \cdot dy \cdot dz$$

With the analyses of the other spacial directions, with addition, rearranging and division by $\rho \cdot dx \cdot dy \cdot dz$ we receive:

$$0 = \frac{\partial c_x}{\partial x} + \frac{\partial c_y}{\partial y} + \frac{\partial c_z}{\partial z}$$

This is the continuity equation for an incompressible fluid in Cartesian coordinates. No characteristic dimensionless number can derived from this equation, as it does not include any specific parameters of a fluid or any geometric features. The continuity equation is the first one of our required five equations.

3.1.2   Equation of motion

This equation results from a force balance on a mass element in a flowing fluid. In the following section, s is a generalized space coordinate.

$$\sum dF_{outer} = dm \cdot \frac{dc_x}{dt} = \rho \cdot dV \cdot \frac{dc_x}{dt}$$

This balance shows that the temporal change of the velocity, and the momentum respectively, is caused by an outer force. If forces caused by electric, magnetic or gravitational fields are excluded, then the outer forces are only resulting from pressure and viscous forces. The last one can be represented by the sheer stress τ, and we receive:

$$\sum dF_{outer} = -dp \cdot dA + \sum d\tau \cdot dA$$

Further, when we take into account, that $dV = dA \cdot ds$ and $c_x = f(s,t)$, i.e. the velocity is a general function of space and time, we can write:

$$\rho \cdot \frac{dc_x}{dt} = \rho \cdot \left(\frac{\partial c_x}{\partial t} + \frac{\partial c_x}{\partial s} \frac{\partial s}{\partial t} \right) = -\frac{\partial p}{\partial s} + \sum \frac{\partial \tau}{\partial s}$$

The sheer stress is given by *Newton*'s law as $\tau = \eta \cdot dc_x/ds$. With this we receive for a steady state flow:

$$\rho \cdot \left(\frac{\partial c_x}{\partial s} c_x \right) = -\frac{\partial p}{\partial s} + \eta \cdot \sum \frac{\partial^2 c_x}{\partial s^2}$$

or, for a flow with a single velocity component in x direction, $c_x = f(x,y)$:

$$\rho \cdot \frac{\partial c_x}{\partial x} c_x = -\frac{\partial p}{\partial x} + \eta \cdot \sum \frac{\partial^2 c_x}{\partial y^2}$$

With some manipulation of this equation, all parameters representing fluid properties, state and geometry of the specific flow, can be condensed into one number. The usual way to start the manipulation is to transfer it in a dimensionless form. This is done by correlating all variables to characteristic parameters.

- correlated length L: $x = x^* \cdot L$
- correlated velocity c: $c_x = c_x^* \cdot c$
- correlated pressure $\rho \cdot c^2$: $p = p^* \cdot \rho \cdot c^2$

The variables with *, x^*, c_x^* and p^* are therefore dimensionless spacial coordinate, dimensionless velocity component and dimensionless pressure. Later on, we will see which correlation parameters L, c can be applied. Inserting the dimensionless variables into the equation of motion we receive:

$$c_x^* \frac{\partial c_x^*}{\partial x^*} = -\frac{\partial p^*}{\partial x^*} + \frac{\eta}{\rho \cdot c \cdot L} \cdot \sum \frac{\partial^2 c_x^*}{\partial y^{*2}}$$

The reciprocal in front of the differential in the third term is called *Reynolds number*. It is the ratio of the force of inertia to the frictional force.

$$Re_L = \frac{c \cdot L}{v} = \frac{c \cdot L \cdot \rho}{\eta} = \frac{\dot{m} \cdot L}{A \cdot \eta} \tag{3.4}$$

The mean velocity of the flow is c, the characteristic length L, the kinematic viscosity of the fluid v and η is the dynamic viscosity. Usually the *Reynolds* number has an index representing the characteristic length L. All problem-specific parameters in the motion equation are condensed into *Re*. Therefore, for the same *Reynolds* number we will always receive identical results for the dependent variables. Conversely, we can conclude: Different solutions of the differential equation for different fluids, state or geometry are caused only by different *Reynolds* numbers. With building equations of motion for the other space directions we receive three additional equations for the five unknown variables, i.e. we need only one more, which will be introduced now.

3.1.3   Equation of energy

The *Reynolds* number contains the velocity, viscosity and the characteristic length. From the previous paragraphs we know that for the heat transfer further parameters, like the thermal conductivity and heat capacity, are relevant. Therefore we will derive the equation for energy, introduce dimensionless variables and try to identify further characteristic dimensionless numbers.

$$d\dot{m}_x = -\frac{\partial c_x}{\partial x} \rho \cdot dx \cdot dy \cdot dz$$

We start from the differential equation of the temperature distribution in a static fluid as derived from an energy balance in Chapter 2.2.1:

$$\frac{\partial \vartheta}{\partial t} = a \cdot \left(\frac{\partial^2 \vartheta}{\partial x^2} + \frac{\partial^2 \vartheta}{\partial y^2} + \frac{\partial^2 \vartheta}{\partial z^2} \right)$$

For a flowing fluid on the left-hand side the temperature change due to enthalpy transfer must be added. Concerning this, we replace the partial differential of the temperature by the total differential and consider that $dx/dt = c_x$, $dy/dt = c_y$ and $dz/dt = c_z$. With this we receive:

$$\frac{d\vartheta}{dt} = \frac{\partial\vartheta}{\partial t} + \frac{\partial\vartheta}{\partial x}\frac{dx}{dt} + \frac{\partial\vartheta}{\partial y}\frac{dy}{dt} + \frac{\partial\vartheta}{\partial z}\frac{dz}{dt} =$$

$$\frac{\partial\vartheta}{\partial t} + \frac{\partial\vartheta}{\partial x}c_x + \frac{\partial\vartheta}{\partial y}c_y + \frac{\partial\vartheta}{\partial z}c_z = a \cdot \left(\frac{\partial^2\vartheta}{\partial x^2} + \frac{\partial^2\vartheta}{\partial y^2} + \frac{\partial^2\vartheta}{\partial z^2} \right)$$

The variables in this equation can be replaced by dimensionless variables. The parameters required in addition are:

- correlated temperature difference $(\vartheta_F - \vartheta_W)$: $\vartheta = \vartheta^* \cdot (\vartheta_F - \vartheta_W) + \vartheta_F$
- correlated time L/c: $t = t^* \cdot L/c$

With inserting the dimensionless variables and rearranging, we receive the following dimensionless differential equation for the temperature distribution:

$$\frac{\partial\vartheta^*}{\partial t^*} + \frac{\partial\vartheta^*}{\partial x^*}c_x^* + \frac{\partial\vartheta^*}{\partial y^*}c_y^* + \frac{\partial\vartheta^*}{\partial z^*}c_z^* = \frac{a}{c \cdot L} \cdot \left(\frac{\partial^2\vartheta^*}{\partial x^{*2}} + \frac{\partial^2\vartheta^*}{\partial y^{*2}} + \frac{\partial^2\vartheta^*}{\partial z^{*2}} \right)$$

The first term on the right-hand side of the equation can be rearranged as follows:

$$\frac{a}{c \cdot L} = \frac{v}{c \cdot L} \cdot \frac{a}{v} = \frac{1}{Re} \cdot \frac{1}{Pr}$$

Beside the *Reynolds* number, a new dimensionless number occurs, called *Prandtl* number.

$$Pr = \frac{v}{a} = \frac{v \cdot \rho \cdot c_p}{\lambda} \tag{3.5}$$

It can be assumed as the ratio of the expansion of the laminar flow boundary layer to the expansion of the temperature boundary layer. Gases have a *Prandtl* number of approximately 0.7, that of the liquids variates in a larger range and is mainly dependent on the temperature.

We have now derived five equations for five dependent variables. From those equations, we extracted two characteristic dimensionless numbers for the convective heat transfer. Now we bring Equation (3.2) in a dimensionless form:

$$\frac{\alpha \cdot L}{\lambda} = -\left(\frac{\partial\vartheta^*}{\partial r^*} \right)_{r^*=1} = Nu_L \tag{3.6}$$

The dimensionless heat transfer coefficient is called *Nußelt number*. It is the ratio of the characteristic length L and the thickness of temperature boundary layer δ_ϑ.

At the same time the derivation shows that the dimensionless heat transfer coefficient only depends on the dimensionless temperature distribution! The latter we receive as a solution of the system of equations derived above. This solution however, is characterized by the *Reynolds* and *Prandtl* number and by the flow geometry. Therefore, *Nußelt* numbers can generally be given in the following form:

$$Nu_L = f(Re_L, Pr, \text{Geometry}, \vartheta / \vartheta_w) \tag{3.7}$$

The last term takes into account the direction of the heat flow rate. This particularity will be discussed later. *Nußelt* numbers in the form mentioned above were experimentally determined for different geometries, fluids and flow conditions. Functions which represent the experimental data with best accuracy were published from many authors. One of the most comprehensive presentations of *Nußelt* functions are presented in VDI-Heat Atlas [3.4].

> *The determination of the heat transfer coefficients can be reduced to the determination of the Nußelt number appropriate for the actual problem. The heat transfer coefficient is then calculated according to the above given definition for Nu.*

The following chapters will describe this procedure for several technically important applications and will illustrate it by numerous examples.

3.2 Determination of heat transfer coefficients

As already mentioned, the *Nußelt* number is a function of *Reynolds* number, material properties, geometry and the direction of heat flux. The influence of material properties is considered by the *Prandtl* number. The heat transfer coefficient will be determined with the *Nußelt* number as defined by Equation (3.4).

3.2.1 Flow in a circular tube

For *flows in circular tubes* the characteristic length is the internal diameter d_i of the tube. The *Nußelt* numbers for laminar flow can be analytically determined. For turbulent flows empirical correlations were found. Laminar and turbulent heat transfer coefficient will be discussed separately.

3.2.1.1 Turbulent flow in circular tubes

The temperature profile in *turbulent flow in a circular* tube is shown in Figure 3.1. The temperature of the fluid is assumed as the temperature in the center of the tube. The state of the art empirical correlation for the *Nußelt* number, which has best fit with experimental results, is [3.3]:

$$Nu_{d_i,turb} = \frac{(\xi/8) \cdot Re_{d_i} \cdot Pr}{1 + 12.7 \cdot \sqrt{\xi/8} \cdot (Pr^{2/3} - 1)} \cdot f_1 \cdot f_2 \tag{3.8}$$

Therein ξ is the *hydraulic friction factor*:

$$\xi = [1.8 \cdot \log(Re_{d_i}) - 1,5]^{-2} \tag{3.9}$$

The material properties have to be determined with the temperature of the fluid in the center of the tube.

Equation (3.8) demohstrate the fact that there is a fundamental correlation between heat transfer and hydraulic friction. The higher the hydraulic friction factor ξ, the higher the *Nußelt* number and the heat transfer coefficient. The higher hydraulic friction requires more power to maintain the flow and the higher heat transfer coefficient results in a smaller heat exchanger. Therefore, the engineers' task is to find an optimum solution with regard to power consumption and heat exchanger size.

The function f_1 considers the influence of tube length and f_2 that of the direction of the heat flux. The tube length influences the heat transfer coefficient because at the tube inlet the temperature profile and the boundary layer are not developed. At the inlet the thickness of the boundary layer is zero and subsequently the heat transfer is infinite.

The thickness of the thermal boundary layer increases with the tube length and the local heat transfer coefficients decrease until the thermal boundary layer is fully developed. Usually for the design of heat exchangers not the local but the mean heat transfer coefficient of the entire tube is of interest. The higher heat transfer coefficients at the tube inlet increases the mean heat transfer coefficient. The integration over the tube length results in function f_1, which considers the *influence of the tube length* on the mean heat transfer coefficient.

$$f_1 = \frac{1}{l} \cdot \int_0^l \left[1 + \frac{1}{3} (d_i / x)^{2/3} \right] \cdot dx = 1 + (d_i / l)^{2/3} \tag{3.10}$$

The term in square brackets is the function for the local heat transfer coefficients.

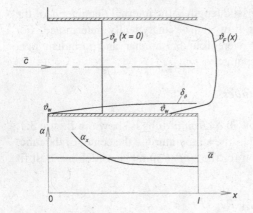

Figure 3.2: Influence of tube length on heat transfer coefficient

The *direction of heat flux* (heating or cooling) influences the heat transfer coefficients because the temperature in the boundary layer is different from that in the tube center, which is used for determination of the *Reynolds* and *Prandtl* number with the temperature dependent material properties. For the function f_2 two different equations are proposed, one for gases and one for liquids.

$$f_2 = \begin{vmatrix} (Pr / Pr_W)^{0.11} & \text{for liquids} \\ (T / T_W)^{0.45} & \text{for gases} \end{vmatrix} \tag{3.11}$$

The validity range of Equations (3.8) to (3.11) is:

$$10^4 < Re_{d_i} < 10^6$$

$$l / d_i > 1$$

In heat exchanger tubes the temperatures of the fluid and the wall are not constant. The fluid properties have to be determined with the mean temperature of the fluid in the tube $\vartheta_m = (\vartheta_{in} + \vartheta_{out}) / 2$. Assuming a constant wall temperature the heat rate can be calculated with the log mean temperature difference as given with Equation (1.15). It is calculated with the temperatures at the inlet and outlet of the tube and the temperature of the wall.

$$\dot{Q} = \alpha \cdot A \cdot \Delta\vartheta_m \tag{3.12}$$

In the case of a parallel-flow or counterflow outside the tubes, the heat rate can be calculated with the overall heat transfer coefficient and the log mean temperature difference. The mean wall temperature can be estimated as:

$$\vartheta_{Wi} = \vartheta_{mi} + \frac{k \cdot d_i}{\alpha_i \cdot d_a} \cdot \Delta\vartheta_m \qquad \vartheta_{Wa} = \vartheta_{ma} - \frac{k}{\alpha_a} \cdot \Delta\vartheta_m \tag{3.13}$$

For rough estimates, instead of Equation (3.8) a simplified exponential equation, which allows an accuracy of 5 %, can be used.

$$Nu_{d_i} = 0.0235 \cdot (Re_{d_i}^{0.8} - 230) \cdot Pr^{0.48} \cdot f_1 \cdot f_2 \tag{3.14}$$

To consider the influence of the direction of heat flux some additional correlations are published in VDI-Heat Atlas [3.4].

3.2.1.2   *Laminar flow in circular tubes at constant wall temperature*

In this book only the case of laminar heat transfer at constant wall temperature is discussed. In [3.4] correlations for heat transfer with constant heat flux are published.

At laminar flow in very long tubes (thermally and hydraulically completely developed temperature profile) the heat transfer coefficient is independent of *Reynolds*- and *Prandtl* number. The *Nußelt* number then has a constant value.

$$Nu_{d_i,lam} = 3.66 \tag{3.15}$$

In shorter tubes the boundary layer is not yet developed and the following relationship was analytically determined:

$$Nu_{d_i,lam} = 0.644 \cdot \sqrt[3]{Pr} \cdot \sqrt{Re_{d_i} \cdot d_i / l} \tag{3.16}$$

With increasing tube length the cross-over to fully developed flows is asymptotic and the *Nußelt* number can be determined with the following equation:

$$Nu_{d_i,lam} = \sqrt[3]{3.66^3 + 0.644^3 \cdot Pr \cdot (Re_{d_i} \cdot d_i / l)^{3/2}} \tag{3.17}$$

Equation (3.17) is valid for *Reynolds* numbers below 2300. Figure 3.3 shows the *Nußelt* numbers for *Pr* = 1 versus *Reynolds* number for different tube lengths.

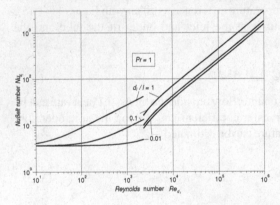

Figure 3.3: Jumps of the *Nußelt* number at the transition from laminar to turbulent

3.2.1.3   Equations for the transition from laminar to turbulent

As shown in diagram in Figure 3.3 at the transition from laminar to turbulent flow there are unsteady jumps in the functions for the *Nußelt* numbers. These jumps cannot be observed in a single experiment, but are the result of correlating the results of many different experiments. The flow in the region between Re = 2300 and Re = 10^4 is called transitional flow, i.e. it is in a transition between laminar and fully turbulent. Equation (3.8) is always valid for *Re* > 10^4 and only below Re = 2300 the flow is laminar. For the transition zone 2300 < Re_{di} < 10^4 following interpolation has been proposed by Gnielinski [3.4]:

$$Nu_{d_i} = (1-\gamma) \cdot Nu_{d_i,lam}(Re = 2300) + \gamma \cdot Nu_{d_i,turb}(Re = 10^4)$$

$$\text{with } \gamma = \frac{Re - 2300}{7700} \tag{3.18}$$

Figure 3.4 shows the interpolated *Nußelt* numbers.

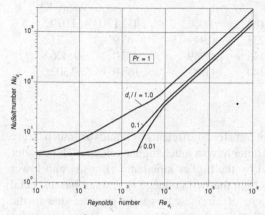

Figure 3.4: *Nußelt* versus *Reynolds* number for different d_i/l at $Pr = 1$

EXAMPLE 3.1: Heat transfer coefficient in a circular pipe

To demonstrate the magnitude of heat transfer coefficients of different fluids they shall be determined in a tube with 25 mm internal diameter. The temperature of the tube wall is 90 °C, that of the fluid 50 °C. Velocities and material properties are listed below:

		velocity m/s	kin. viscosity 10^{-6} m²/s	thermal conductivity W/(m K)	Pr	Pr_{Wi}
Water		2	0.554	0.6410	3.570	1.96
Air	1 bar	20	18.250	0.0279	0.711	
Air	10 bar	20	1.833	0.0283	0.712	
R134a	10 bar	2	0.146	0.0751	3.130	3.13

Find: The heat transfer coefficients

Solution

Assumptions

- Temperature of tube wall and fluid are constant.
- The influence of the tube length can be neglected.

Analysis

As will be seen, the *Reynolds* number is always grater than 10^4, therefore all heat transfer coefficients can be calculated with Equation (3.8).

	Re_{di}	ξ	f_2	$Nu_{di,turb}$	α $W/(m^2\,K)$
Water	90 253	0.0182	1.068	432.1	11 079.5
Air 1 bar	27 397	0.0238	0.949	63.9	71.3
Air 10 bar	272 777	0.0146	0.949	377.0	426.8
R134a	342 466	0.0140	1.000	1 166.0	3 502.6

Discussion

The calculations show that the heat transfer coefficients of liquids are much larger than those of the gases, although the latter have a much higher velocity. The smaller heat transfer coefficients are caused by the higher kinematic viscosity and lower thermal conductivity. Because of its high thermal conductivity, water has a special position. The heat transfer coefficient of air increases with pressure due to the decrease of kinematic viscosity. The influence of the tube was neglected as it is not relevant for the comparison of different material properties.

EXAMPLE 3.2: Heat transfer coefficients of a heat exchanger

In a heat exchanger with tubes of 1 m length, 15 mm outer diameter and 1 mm wall thickness water flows with 1 m/s velocity. On the outer wall of the tubes, Freon R134a condenses at 50 °C. The freon heat transfer coefficient is 5 500 $W/(m^2\,K)$. The thermal conductivity of the tube material is 230 $W/(m\,K)$. The water enters the tubes with a temperature of 20 °C.

Material properties of water:

	ρ kg/m^3	c_p J/(kg K)	λ W/(m K)	ν m^2/s	Pr -
20 °C:	998.2	4 184	0.598	$1.003 \cdot 10^{-6}$	7.00
30 °C:	995.7	4 180	0.616	$0.801 \cdot 10^{-6}$	5.41
40 °C:	992.3	4 178	0.631	$0.658 \cdot 10^{-6}$	4.32

Find

The heat transfer coefficient of the water flow, the outlet temperature of the water and the heat rate.

Solution

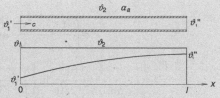

Schematic   See sketch.

Assumption

- The mean heat transfer coefficient is constant.

Analysis

The outlet temperature of the water is not known, therefore the mean temperature required for the determination of the material properties must be assumed. The outlet temperature can be determined when the heat transfer coefficient and heat rate are calculated. Initially an outlet temperature of 30 °C is assumed. The mean water temperature then is 25 °C. The interpolated material properties of water are:

$\rho = 997.0\,\text{kg/m}^3, c_p\,4182\,\text{J/(kg K)}, \lambda = 0.607\,\text{W/(m K)}, Pr = 6.21, \nu = 0.902 \cdot 10^{-6}\,\text{m}^2/\text{s}.$

The *Reynolds* number is: $Re = \dfrac{c \cdot d_i}{\nu} = \dfrac{1 \cdot \text{m} \cdot 0.013 \cdot \text{m} \cdot \text{s}}{\text{s} \cdot 0.902 \cdot 10^{-6} \cdot \text{m}^2} = 14412$

Hydraulic friction factor (3.9): $\xi = [1.8 \cdot \log(Re_{d_i}) - 1.5]^{-2} = 0.0279$

The *Nußelt* number can now be calculated with Equations (3.8), (3.10) and (3.11). To determine the influence of the direction of heat flux with Equation (3.11) the *Prandtl* number at wall temperature is required. As for its determination, the overall heat transfer coefficient must be known and iteration must be performed. The value of f_2 is initially assumed as 1. The function f_1 is:

$$f_1 = 1 + (d_i / l)^{2/3} = 1 + (0.013 \cdot \text{m} / 1)^{2/3} = 1.055$$

The *Nußelt* number calculated with Equation (3.8):

$$Nu_{d_i,turb} = \frac{(\xi / 8) \cdot Re_{d_i} \cdot Pr}{1 + 12.7 \cdot \sqrt{\xi / 8} \cdot (Pr^{2/3} - 1)} \cdot f_1 \cdot f_2 =$$

$$= \frac{0,0035 \cdot 14\,412 \cdot 6.21}{1 + 12.7 \cdot \sqrt{0.0035} \cdot (6.21^{2/3} - 1)} \cdot 1.055 = 118.3$$

For the heat transfer coefficient we receive:

$$\alpha = Nu_{d_i} \cdot \lambda / d_i = 118.3 \cdot 0.607 \cdot \text{W/(m} \cdot \text{K)}/(0.013 \cdot \text{m}) = 5524\,\text{W/(m}^2 \cdot \text{K)}$$

The overall heat transfer coefficient calculated with Equation (2.27) is:

$$k = \left(\frac{1}{\alpha_a} + \frac{d_a}{2 \cdot \lambda_R} \cdot \ln \frac{d_a}{d_i} + \frac{d_a}{d_i \cdot \alpha_i} \right)^{-1} =$$

$$= \left(\frac{1}{5500} + \frac{0.015}{2 \cdot 230} \cdot \ln \frac{15}{13} + \frac{15}{13 \cdot 5524} \right)^{-1} = 2529 \ \frac{W}{m^2 \cdot K}$$

With the overall heat transfer coefficient and the log mean temperature the wall temperature can be determined. The log mean temperature is:

$$\Delta \vartheta_m = \frac{\vartheta_1'' - \vartheta_1'}{\ln \left(\dfrac{\vartheta_2 - \vartheta_1'}{\vartheta_2 - \vartheta_1''} \right)} = \frac{(30 - 20) \cdot K}{\ln \left(\dfrac{50 - 20}{50 - 30} \right)} = 24.66 \ K$$

The wall temperature according to Equation (3.13) is:

$$\vartheta_W = \vartheta_m + \Delta \vartheta_m \cdot \frac{k \cdot d_i}{\alpha_i \cdot d_a} = 25 \ °C + 24.66 \cdot K \cdot \frac{2529 \cdot 13}{5524 \cdot 15} = 34.8 \ °C$$

The linearly interpolated *Prandtl* number at 34.8 °C is 4.89.

Equation (3.11) delivers for f_2: $f_2 = (Pr / Pr_W)^{0.11} = (6.21 / 4.89)^{0.11} = 1.027$

The *Nußelt* number and also the heat transfer coefficient will be 3 % larger. For α_i, k and ϑ_W the following values were determined: $\alpha_i = 5671$ W/(m²K), $k = 2565$ W/(m² K), $\vartheta_W = 34.7$ °C. Pr_W is then 4.901 and $f_2 = 1.0263$. With these values α_i and k can be determined again:

$$\alpha_i = 5670 \ W/(m^2 \ K) \qquad k = 2564 \ W/(m^2 \ K)$$

With the rate equation the heat rate can be determined.

$$\dot{Q} = k \cdot A \cdot \Delta \vartheta_m = k \cdot \pi \cdot d_a \cdot l \cdot \Delta \vartheta_m =$$

$$= 2564 \cdot W/(m^2 \cdot K) \cdot \pi \cdot 0.015 \cdot m \cdot 1 \cdot m \cdot 24.66 \cdot K = 2980 \ W$$

With the energy balance equation the water outlet temperature is calculated. First the mass flow rate of water in the pipe has to be determined.

$$\dot{m} = c \cdot 0.25 \cdot \pi \cdot d_i^2 \cdot \rho = 1 \cdot m/s \cdot 0.25 \cdot \pi \cdot 0.013^2 \cdot m^2 \cdot 997 \cdot kg/s = 0.132 \ kg/s$$

$$\vartheta_1'' = \vartheta_1' + \frac{\dot{Q}}{\dot{m} \cdot c_p} = 20 \ °C + \frac{2980 \cdot W}{0.132 \cdot kg/s \cdot 4182 \cdot J/(kg \cdot K)} = 25.4 \ °C$$

The mean temperature of the water is not as assumed 25 °C but 22.7 °C. The whole calculation procedure must be repeated with the following material properties:

$\rho = 997.5$ kg/m³, c_p 4182 J/(kg K), $\lambda = 0.607$ W/(m K), $\nu = 0.948 \cdot 10^{-6}$ m²/s, $Pr = 6.53$.

The calculations will not be shown in detail, only the results are listed below.

Re_{di}	α_i	k	$\Delta\vartheta_m$	ϑ_w	f_2	$\dot{Q}$	ϑ''_1
	W/(m² K)	W/(m² K)	K	°C		kW	°C
13706	5 525	2 530	27.22	33.49	1.0297		
	5 544	2 534	27.22	33.47	1.0298		
	5 543	**2 534**				**3.251**	**25.87**
13778	5 559	2 538	26.96	33.60	1.0295		
	5 557	2 537	26.96	33.60	1.0295		
	5 557	**2 537**				**3.223**	**25.82**

As the differences of the last values are below 0.2 %, the iteration was terminated.

Discussion

Iteration procedures are generally necessary for the determination of heat transfer coefficients of heat exchangers. In this example the iteration could have been terminated after the first run. The heat rate was already determined with 1 % accuracy. Normally computer codes are developed for the design of heat exchangers, in which the heat transfer equations and material properties are programmed. In many cases it is sufficient to give the material properties at two temperatures and use linear interpolation. Anyhow, always it must be checked if the required accuracy is reached.

EXAMPLE 3.3: Design of a power plant condenser

In a power plant condenser a heat rate of 2 000 MW has to be transferred to the cooling water. The condenser is equipped with titanium tubes of 24 mm outer diameter and 0.5 mm wall thickness. The cooling water velocity is 2 m/s. Titanium has a thermal conductivity of 16 W/(m K). The steam condenses at the saturation temperature of 35 °C, the heat transfer of condensation outside the tubes is given as 13 500 W/(m² K). The cooling water is heated from 20 °C to 30 °C. At 25 °C the water has following material properties:
$\rho = 997.0$ kg/m³, $c_p = 4182$ J/(kg K), $Pr = 6.2$, $\lambda = 0.607$ W/(m K),
$\nu = 0.902 \cdot 10^{-6}$ m²/s.
The influence of the direction of heat flux can be neglected.

Find

a) the required number of tubes
b) the required tube' length
c) the condensation temperature, if the heat transfer coefficient drops 10 % due to
   fouling.

Solution

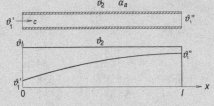

Schematic   See sketch.

Assumptions

- The mean heat transfer coefficient is constant.
- The influence of the direction of the heat flux can be neglected, i.e. $f_2 = 1$.

Analysis

a)   With the energy balance equation the mass flow rate of the water can be
determined.

$$\dot{m} = \frac{\dot{Q}}{c_p \cdot (\vartheta_1'' - \vartheta_1')} = \frac{2\,000 \cdot 10^6 \cdot W \cdot kg \cdot K}{4182 \cdot J \cdot (30 - 20) \cdot K} = 47\,824\ \frac{kg}{s}$$

As the water velocity is given, the mass flow rate in one tube can be calculated.

$$\dot{m}_{1tube} = c \cdot 0.25 \cdot \pi \cdot d_i^2 \cdot \rho = 2 \cdot m/s \cdot 0.25 \cdot \pi \cdot 0.023^2 \cdot m^2 \cdot 997 \cdot kg/s = 0.828\ kg/s$$

To have a mass flow rate of 47 824 kg/s **57 727** tubes are required.

b)   To calculate the *Nußelt* number with Equation (3.8) the function f_1 is needed.
It takes into account the influence of the tube length, which is not yet known. For the
first calculation it is assumed: $f_1 = 1$.

Reynolds number: $Re = \dfrac{c \cdot d_i}{\nu} = \dfrac{2 \cdot m \cdot 0.023 \cdot m \cdot s}{s \cdot 0.902 \cdot 10^{-6} \cdot m^2} = 50\,998$

Hydraulic friction factor (3.9): $\xi = [1.8 \cdot \log(Re_{d_i}) - 1.5]^{-2} = 0.0206$

$$Nu_{d_i,turb} = \frac{(\xi/8) \cdot Re_{d_i} \cdot Pr}{1 + 12.7 \cdot \sqrt{\xi/8} \cdot (Pr^{2/3} - 1)} = \frac{0.00257 \cdot 50998 \cdot 6.2}{1 + 12.7 \cdot \sqrt{0.00257} \cdot (6.2^{2/3} - 1)} = 321.3$$

For the heat transfer coefficient in the tubes we receive:

$$\alpha_i = Nu_{di} \cdot \lambda / d_i = 321.3 \cdot 0.607 \cdot W/(m \cdot K)/(0.023 \cdot m) = 8481 \ W/(m^2 \cdot K)$$

and for the overall heat transfer coefficient with Equation (2.27):

$$k = \left(\frac{1}{\alpha_a} + \frac{d_a}{2 \cdot \lambda_R} \cdot \ln \frac{d_a}{d_i} + \frac{d_a}{d_i \cdot \alpha_i} \right)^{-1} = \left(\frac{1}{13500} + \frac{0.024}{2 \cdot 16} \cdot \ln \frac{24}{23} + \frac{24}{23 \cdot 8481} \right)^{-1} =$$
$$= 4366 \ W/(m^2 \cdot K)$$

With the overall heat transfer coefficient and the log mean temperature difference the required heat transfer surface area can be determined. The log mean temperature difference is:

$$\Delta \vartheta_m = \frac{\vartheta_1'' - \vartheta_1'}{\ln \left(\dfrac{\vartheta_2 - \vartheta_1'}{\vartheta_2 - \vartheta_1''} \right)} = \frac{(30 - 20) \cdot K}{\ln \left(\dfrac{35 - 20}{35 - 30} \right)} = 9.102 \ K$$

The required surface area is calculated with the rate equation.

$$A = \frac{\dot{Q}}{k \cdot \Delta \vartheta_m} = \frac{2\,000 \cdot 10^6 \cdot W \cdot m^2 \cdot K}{4366 \cdot W \cdot 9.102 \cdot K} = 50325 \ m^2$$

For this area the following tube length is required:

$$l = \frac{A}{n \cdot \pi \cdot d_a} = \frac{50325 \cdot m^2}{57727 \cdot \pi \cdot 0.024 \cdot m} = 11.562 \ m$$

With this tube length we receive for the function $f_1 = 1.016$. Therewith the transfer coefficient is 1.6 % higher and the overall heat transfer coefficient increases 1.43 %, the tube length decreases correspondingly. We receive it with 11.460 m. The next iteration delivers **11.465 m**.

c) The flow rate of the steam to the condenser will not change with fouling, thus the heat rate and also the cooling water outlet temperature remain the same as with the clean condenser tubes. To be able to maintain the heat rate, according to the rate equation with the reduced overall heat transfer coefficient the log mean temperature difference must increase correspondingly. As the cooling water temperatures do not change, the condensation temperature will increase. The log mean temperature difference with the "fouled" overall heat transfer k_v coefficient is:

$$\Delta \vartheta_m = \frac{\dot{Q}}{A \cdot k_v} = \frac{2 \cdot 10^9 \cdot W \cdot m^2 \cdot K}{49901 \cdot m^2 \cdot 4403 \cdot 0.9 \cdot W} = 10.114 \ K$$

With Equation (1.15) the changed condensation temperature can be determined.

$$\vartheta_2 = \frac{\vartheta_1' - \vartheta_1'' \cdot e^{\frac{\vartheta_1' - \vartheta_1'}{\Delta\vartheta_m}}}{1 - e^{\frac{\vartheta_1' - \vartheta_1'}{\Delta\vartheta_m}}} = 35.93\,°C$$

Discussion

At the stage of condenser design the tube length has to be determined, therefore the function f_1 is not known and an iteration is necessary.

The condensation temperature increases by the reduced overall heat transfer coefficient, caused by fouling. To be able to keep the heat rate with the reduced overall heat transfer coefficient, the log mean temperature difference must increase. As the cooling water temperature does not change, the condensation temperature must increase. This is a real problem in power plants, as with the increased condensation temperature the condenser pressure is increasing and consequently the power output decreases. In this example the condenser pressure increases from 56 to 59 mbar. This decreases the power output by 0.053 %, which in the case of such a large condenser power, is approximately 5.3 MW.

3.2.1.3   *Flow in tubes and channels of non-circular cross-sections*

In tubes and channels of *non-circular cross-section* for turbulent flows the *Nußelt* number can be calculated with the same equations as used for circular tubes. Instead of the tube diameter the *characteristic length* will be the *hydraulic diameter* of the flow channel, with which the *Reynolds* and *Nußelt* number can be determined.

The hydraulic diameter is defined as:

$$d_h = \frac{4 \cdot A_{CS}}{U} = \frac{4 \cdot \text{flow cross-sectional area}}{\text{wetted circumference}} \tag{3.19}$$

For laminar flows in channels of non-circular cross-section the equations used for circular tubes are not valid. For a few simple geometries solutions for the heat transfer coefficients can be found in literature.

For *annuli* (Figure 3.5) with turbulent flow a further correction is required. The ratio of annulus diameter has an influence on heat transfer. For annuli where heat transfer occurs only to or from the inner tube, the following correction is proposed [3.4, 3.5]:

$$Nu_{Annulus} / Nu_{d_h} = 0.86 \cdot (D / d)^{0.16} \tag{3.20}$$

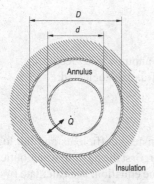

D

d

Annulus

$\dot{Q}$

Insulation

Figure 3.5: Annulus

EXAMPLE 3.4: Design of a counterflow heat exchanger

The heat exchanger of a district heating system in a house consists of an inner tube with 18 mm outer diameter and 1 mm wall thickness. It is installed in the center of an outer tube with 24 mm inner diameter. The flow velocity as well in the inner tube and in the annulus is 1 m/s. The annulus is entered by the heating water with 90 °C. The service water in the tube should be heated from 40 °C to 60 °C. The thermal conductivity of the inner tube material is 17 W/(m K). The outer tube is ideally insulated to the environment. To simplify calculation the functions f_1 and f_2 shall have the value of 1. The material properties are:

	density kg/m³	kin. viscosity 10^{-6} m²/s	heat conductivity W/(m K)	Pr	c_p J/(kg K)
Service water	998.1	0.553	0.6437	3.55	4179
Heating water	971.8	0.365	0.6701	2.22	4195

Find

The required length of the heat exchanger.

Solution

Schematic   See sketch.

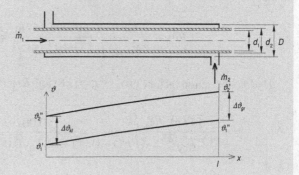

Assumptions

- No heat transfer to the environment.
- The perpendicular flow at the in- and outlet of the annulus is neglected.
- The influence of tube length and direction of heat flux are neglected.

Analysis

To calculate the required surface area the heat transfer coefficients and the log mean temperature difference are required. For the latter, first the heating water outlet temperature must be determined with the energy balance equation. We calculate first the mass flow rate in the inner tube and in the annulus.

$$\dot{m}_1 = c_1 \cdot 0.25 \cdot \pi \cdot d_1^2 \cdot \rho_1 = 1 \cdot \text{m/s} \cdot 0.25 \cdot \pi \cdot 0.016^2 \text{m}^2 \cdot 998.1 \cdot \text{kg/m}^3 = 0.2007 \ \text{kg/s}$$

$$\dot{m}_2 = c_2 \cdot 0.25 \cdot \pi \cdot (D^2 - d_2^2) \cdot \rho_1 = 1 \cdot 0.25 \cdot \pi \cdot (0.024^2 - 0.018^2) \cdot 971.8 = 0.1923 \ \text{kg/s}$$

With the energy balance equation the heat rate to the service water can be determined.

$$\dot{Q} = \dot{m}_1 \cdot c_{p1} \cdot (\vartheta_1'' - \vartheta_1') = 0.2007 \cdot \text{kg/s} \cdot 4179 \cdot \text{J/(kg} \cdot \text{K)} \cdot (60 - 40) \cdot \text{K} = 16.773 \ \text{kW}$$

This heat rate comes from the heating water. The outlet temperature results from the energy balance equation.

$$\vartheta_2'' = \vartheta_2' - \frac{\dot{Q}}{\dot{m}_2 \cdot c_{p2}} = 90 \ ^\circ\text{C} - \frac{16\,773 \cdot \text{W}}{0.1923 \cdot \text{kg/s} \cdot 4195 \cdot \text{J/(kg} \cdot \text{K)}} = 69.21 \ ^\circ\text{C}$$

Log mean temperature:

$$\Delta\vartheta_m = \frac{\Delta\vartheta_{gr} - \Delta\vartheta_{kl}}{\ln(\Delta\vartheta_{gr} / \Delta\vartheta_{kl})} = \frac{(30 - 29.21) \cdot \text{K}}{\ln(30 / 29.21)} = 29.60 \ \text{K}$$

The heat transfer coefficient in the tube will be calculated with Equation (3.8) and with $f_1 = f_2 = 1$.

$$Re_{d_1} = \frac{c_1 \cdot d_1}{\nu_1} = \frac{1 \cdot \text{m/s} \cdot 0.016 \cdot \text{m}}{0.553 \cdot 10^{-6} \cdot \text{m}^2 / \text{s}} = 28933$$

Hydraulic friction factor (3.9): $\xi = [1.8 \cdot \log(Re_{d_1}) - 1.5]^{-2} = 0.0234$

$$Nu_{d_1, turb} = \frac{(\xi / 8) \cdot Re_{d_1} \cdot Pr}{1 + 12.7 \cdot \sqrt{\xi / 8} \cdot (Pr^{2/3} - 1)} = \frac{0.00293 \cdot 28933 \cdot 355}{1 + 12.7 \cdot \sqrt{0.00293} \cdot (3.55^{2/3} - 1)} = 157.4$$

Heat transfer coefficient in the inner tube:

$$\alpha_i = Nu_{d_i} \cdot \lambda / d_i = 157.4 \cdot 0.6437 \cdot \text{W/(m} \cdot \text{K)}/(0.016 \cdot \text{m}) = 6333 \text{ W/(m}^2 \cdot \text{K)}$$

To calculate the heat transfer coefficient in the annulus first the hydraulic diameter must be determined with Equation (3.19).

$$d_h = \frac{4 \cdot A_{CS}}{U} = \frac{\pi \cdot (D^2 - d_2^2)}{\pi \cdot (D + d_2)} = D - d_2 = 6 \text{ mm}$$

The heat transfer coefficient can be calculated with Equations (3.8) and (3.20).

$$Re_{d_h} = \frac{c_2 \cdot d_h}{v_1} = \frac{1 \cdot \text{m/s} \cdot 0.006 \cdot \text{m}}{0.365 \cdot 10^{-6} \cdot \text{m}^2/\text{s}} = 16438$$

Hydraulic friction factor (3.9): $\xi = [1.8 \cdot \log(Re_{d_h}) - 1.5]^{-2} = 0.0270$

$$Nu_{d_h,turb} = \frac{(\xi / 8) \cdot Re_{d_h} \cdot Pr}{1 + 12.7 \cdot \sqrt{\xi / 8} \cdot (Pr^{2/3} - 1)} \cdot 0.86 \cdot \left(\frac{D}{d_2}\right)^{0.16} =$$

$$= \frac{0.00337 \cdot 16438 \cdot 2.22}{1 + 12.7 \cdot \sqrt{0.00337} \cdot (2.22^{2/3} - 1)} \cdot 0.86 \cdot \left(\frac{24}{18}\right)^{0.16} = 73.0$$

Heat transfer coefficient in the annulus:

$$\alpha_a = Nu_{d_h} \cdot \lambda_2 / d_h = 73 \cdot 0.6701 \cdot \text{W/(m} \cdot \text{K)}/(0.006 \cdot \text{m}) = 8155 \text{ W/(m}^2 \cdot \text{K)}$$

Overall heat transfer coefficient (2.27):

$$k = \left(\frac{1}{\alpha_a} + \frac{d_2}{2 \cdot \lambda_R} \cdot \ln\frac{d_2}{d_1} + \frac{d_2}{d_1 \cdot \alpha_i}\right)^{-1} =$$

$$= \left(\frac{1}{8155} + \frac{0.018}{2 \cdot 17} \cdot \ln\frac{18}{16} + \frac{18}{16 \cdot 6332}\right)^{-1} = 2758 \; \frac{\text{W}}{\text{m}^2 \cdot \text{K}}$$

The required heat exchanger surface area is calculated with the rate equation:

$$A = \frac{\dot{Q}}{k \cdot \Delta \vartheta_m} = \frac{16773 \cdot \text{W} \cdot \text{m}^2 \cdot \text{K}}{2758 \cdot \text{W} \cdot 29.6 \cdot \text{K}} = 0.205 \text{ m}^2$$

To have the above surface area the following length is required:

$$l = \frac{A}{\pi \cdot d_2} = \textbf{3.63 m}$$

Discussion

In heat exchangers with water flow the heat transfer coefficients are rather high and large heat rates can be transferred in exchanger with small surface areas.

Without the simplifying assumptions to neglect the influence of tube length and the direction of heat flux, iterations would be required with three times the amount of time needed for the calculations. The exact surface area would be 1.0 % smaller.

3.2.2   Flat plate in parallel flow

In technical applications heat transfer of a fluid flowing along a *flat plate* is rather rare. The determination of the heat transfer coefficients is more simple than that of many other bodies and is therefore often discussed in books to demonstrate the correlation between heat transfer and hydraulic resistance. In this book only the corresponding formula will be given. The characteristic length is the length L of the wall in flow direction.

For the laminar flow we receive the same formula as for the tube:

$$Nu_{L,lam} = 0.644 \cdot \sqrt[3]{Pr} \cdot \sqrt{Re_L} \qquad \text{for } Re_L < 10^5 \qquad (3.21)$$

The empiric correlation for the turbulent flow is:

$$Nu_{L,turb} = \frac{0.037 \cdot Re_L^{0.8} \cdot Pr}{1 + 2.443 \cdot Re_L^{-0,1} \cdot (Pr^{2/3} - 1)} \cdot f_3 \qquad \text{for } 5 \cdot 10^5 < Re_L < 10^7 \quad (3.22)$$

As the *Reynolds* number uses the length of the plane plate no further correction function for the length is required. Function f_3 is a correction function taking into account the influence of heat flux direction. It is given as:

$$f_3 = \begin{cases} (Pr / Pr_W)^{0.25} & \text{for liquids} \\ 1 & \text{for gases} \end{cases} \qquad (3.23)$$

The range of *Reynolds* number from 10^5 to $5 \cdot 10^7$ is not covered with the two equations. The transition from laminar to turbulent is asymptotic and the following equation is proposed for the range $10 < Re_L > 10^7$.

$$Nu_L = \sqrt{Nu_{L,lam}^2 + Nu_{L,turb}^2} \qquad \text{for } 10 < Re_L < 10^7 \qquad (3.24)$$

3.2.3 Single bodies in perpendicular cross-flow

In many applications the external *cross-flow* in shell and tube heat exchangers or cylindrical temperature gauges is normal to the axis of the tubes. Tube bundles will be discussed in Chapter 3.2.4, but their laws are based on the laws of single bodies. As is known from fluid mechanics, when a body is hit by a flow at the stagnation point, first a laminar boundary layer develops, which then, depending on the flow velocity and geometry, transits to a turbulent flow. Further on flow detachment and vortex shedding may occur. The flow processes and therefore also the heat transfer phenomena are very complex. Similar to the tube flow, empiric correlations were found to determine the *Nußelt* number as a function of geometry, *Reynolds* and *Prandtl* number.

For the *Reynolds* and *Nußelt* number the *characteristic length* is the so-called *flow length L'*, which is the heat transfer surface area divided by the projected circumference of the body.

$$L' = A / U_{proj} \qquad (3.25)$$

The *projected circumference U_{proj}* is the circumference of the active heat transfer surface area projected in the direction of the flow. For example, for a pipe with perpendicular cross-flow the projected circumference is twice the length of the tube, for a plane plate the width of the plate and for a sphere its circumference.

Figure 3.6 shows the heat transfer surface areas and the projected circumference of a few bodies.

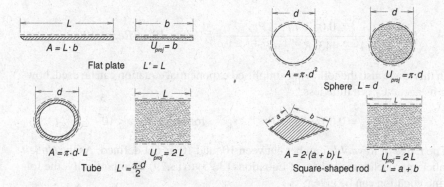

Figure 3.6: Heat transfer surface area and projected circumference of selected bodies

From the surface of a sphere or a cylinder even if the *Reynolds* number reaches zero, heat transfer to the surrounding fluid happens. For a sphere with the inner diameter d, the heat transfer coefficient with an infinite outside diameter (static surroundings) was found in Chapter 2 as (2.36):

$$\alpha_0 = \frac{2 \cdot \lambda}{d} \qquad (3.26)$$

From there the *Nußelt* number is:

$$Nu_{L',0} = 2 \qquad \text{for} \quad Re_{L'} < 0.1 \qquad (3.27)$$

For a cylinder the derivation is not so simple and therefore only the result will be presented here.

$$Nu_{L',0} = 0.3 \qquad \text{for} \quad Re_{L'} < 0.1 \qquad (3.28)$$

For a flat plate at $Re_L = 0$ the *Nußelt* number $Nu_{L',0}$ is zero.

For spheres or cylinders with dimensions smaller than the thickness of the boundary layer at *Reynolds* numbers smaller than 1, the *Nußelt* numbers of spheres and cylinders should be determined using the following formulae:

Sphere: $\qquad Nu_{L',0} = 1.001 \cdot \sqrt[3]{Re_{L'} \cdot Pr} \quad$ for $\; 0.1 < Re_{L'} < 1$

Cylinder: $\qquad Nu_{L',0} = 0.75 \cdot \sqrt[3]{Re_{L'} \cdot Pr} \quad$ for $\; 0.1 < Re_{L'} < 1 \qquad (3.29)$

For *Reynolds* numbers between 1 and 1000 the same equation as for flat plates can be used.

$$Nu_{L',lam} = 0.664 \cdot \sqrt[3]{Pr} \cdot \sqrt{Re_{L'}} \qquad \text{for} \quad 1 < Re_{L'} < 1000 \qquad (3.30)$$

For *Reynolds* between 10^5 and 10^7 the following equation was found [3.4]:

$$Nu_{L',turb} = \frac{0.037 \cdot Re_{L'}^{0.8} \cdot Pr}{1 + 2.443 \cdot Re_{L'}^{-0.1} \cdot (Pr^{2/3} - 1)} \cdot f_4 \qquad \text{for} \quad 10^5 < Re_{L'} < 10^7 \qquad (3.31)$$

In this range also the following simplified exponential equation can be used, however its accuracy is a little less.

$$Nu_{L',turb} = 0.037 \cdot Re_{L'}^{0.8} \cdot Pr^{0.48} \cdot f_4 \qquad \text{for } 10^5 < Re_{L'} < 10^7 \qquad (3.32)$$

The range of *Reynolds* number between 10^3 and 10^5 is not defined. As the *Nußelt* number approaches the values of Equations (3.25) to (3.29) asymptotically, the following equation can be given:

$$Nu_{L'} = Nu_{L',0} + \sqrt{Nu_{L',lam}^2 + Nu_{L',turb}^2} \qquad \text{for} \quad 10 < Re_{L'} < 10^7 \qquad (3.33)$$

The function f_4 considers the influence of the direction of the heat flux.

$$f_4 = \begin{vmatrix} (Pr / Pr_W)^{0.25} & \text{for liquids} \\ (T / T_W)^{0.121} & \text{for gases} \end{vmatrix} \qquad (3.34)$$

Figure 3.7 shows the *Nußelt* number of a cylinder in a perpendicular cross-flow.

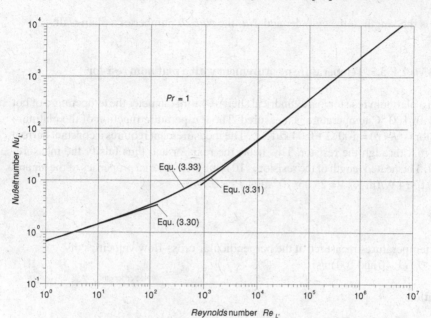

Figure 3.7: *Nußelt* number of cylinder in perpendicular cross-flow

For non-perpendicular cross flow, the heat transfer coefficients will decrease. Figure 3.8 shows the ratio of *Nußelt* number in an angular cross-flow vs. the impact angle.

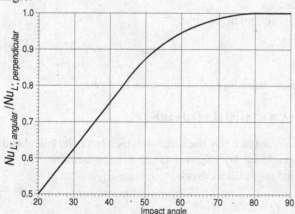

Figure 3.8: Ratio of *Nußelt* number of cylinders with angular to perpendicular cross-flow

A cylinder with parallel flow can be calculated as flat plate. However, if the diameter of the cylinder is close to or smaller as the thickness of the boundary layer, the following equation can be applied.

$$Nu_{L,Zyl} = (1 + 2.3 \cdot (L/d) \cdot Re_L^{-0.5}) \cdot Nu_L \qquad (3.35)$$

L is the length of the cylinder and Nu_L the *Nußelt* number of a plain wall.

EXAMPLE 3.5: Temperature measurement with a platinum resistor

With a platinum resistor in a cylindrical shell of 4 mm diameter the temperature of hot air with 100 °C temperature is measured. The temperature function of the platinum resistor is: $R(\vartheta) = 100\,\Omega + 0.04\,\Omega/\text{K} \cdot \vartheta$. The measurement requires a constant current of 1 mA through the resistor. This heats the resistor and thus falsify the measurement. The heated length of the resistor is 10 mm. The material properties of the air are: $\lambda = 0.0314$ W/(m K), $\nu = 23.06 \cdot 10^{-6}$ m²/s, $Pr = 0.701$.

Find

The temperatures measured at the perpendicular cross-flow velocities of 0.01, 0.1, 1, 10 and 100 m/s

Solution

Schematic   See sketch.

Assumptions

- The temperature in the resistor is constant.
- The air temperature is constant.
- Effects at the end of the resistor are negligible.

Analysis

The current through the resistor produces the following heat rate:

$$\dot{Q} = i^2 \cdot R = i^2 \cdot (100\,\Omega + 0{,}04\,\Omega/\text{K} \cdot \vartheta)$$

The temperature ϑ is the one measured by the resistor, which has to be calculated here. Due to the heat rate produced by the resistance it is higher than the air temperature. Heat is transferred to the air as given by the rate equation:

$$\dot{Q} = \alpha \cdot A \cdot (\vartheta - \vartheta_\infty)$$

The energy conservation requires that the two heat rates have the same value.

$$\vartheta = \frac{i^2 \cdot 100\,\Omega + \alpha \cdot A \cdot \vartheta_\infty}{\alpha \cdot A - i^2 \cdot 0{,}04 \cdot \Omega/\text{K}}$$

For the different flow velocities the *Reynolds* number is calculated with Equation (3.44). The characteristic length is:

$$L' = \frac{A}{U_{proj}} = \frac{\pi \cdot d \cdot l}{2 \cdot l} = \frac{\pi \cdot d}{2} = 6.28 \text{ mm}$$

The results of the calculations are presented in the table below:

c m/s	$Re_{L'}$	$Nu_{L',lam}$	$Nu_{L',turb}$	$Nu_{L'}$	α W/(m² K)	ϑ °C
0.01	2.66	0.962	0.068	1.265	6.320	100.131
0.10	26.60	3.044	0.431	3.374	16.860	100.050
1.00	266.00	9.624	2.719	10.301	51.480	100.016
10.00	2662.00	30.435	17.154	35.236	176.092	100.005
100.00	26624.00	96.244	108.233	145.136	725.311	100.001

Discussion

The current through the resistor heats it up. The error in the measurement at a flow velocity of 0.1 m/s is less then 0.05 K. With state-of -the-art measuring devices the required current can be kept below of 1 mA. In static air the *Nußelt* number is: $Nu_{L',0}$ = 0,3. The value of the heat transfer coefficient is 1.5 W/(m² K) and the deviation 0.5 K. Reduction of the current to 0.1 mA reduces the deviation by 100 times.

3.2.4   Perpendicular cross-flow in tube bundles

In industry, heat exchanger *tube bundles* with perpendicular flow are common practice. Already in a single row of tubes the velocity is higher than with a single tube. Therefore, in the first row of tube bundle the laws for a single tube are already not valid. In the case of multiple tubes rows in the flow direction flow separation and vortex shedding occur. The calculation procedures for tube bundles use the *Nußelt* numbers of the single tubes and correction functions, which take into account the tube arrangements and bundle geometry. Figure 3.9 shows different possibilities of tube arrangements.

The distance of two tubes perpendicular to the flow direction is s_1, the distance of the tube rows s_2. The arrangement of the tubes is characterized by the dimensionless tube distance $a = s_1/d$ and the dimensionless tube row distance $b = s_2/d$.

$$a = s_1/d \tag{3.36}$$

$$b = s_2/d \tag{3.37}$$

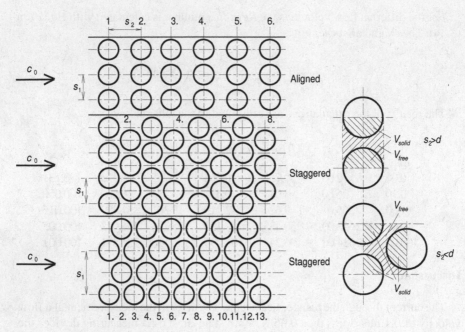

Figure 3.9: Different arrangements of tubes in a tube bundle

The *Reynolds* number is determined with the mean velocity in the hollow (free) space c_ψ. The ratio of the hollow space or *porosity* is ψ. In Figure 3.9 the determination of the hollow volume ratio is presented. Depending on the value of the dimensionless tube row distance two different equations are given for the determination of the hollow volume ratio.

$$\Psi = 1 - \frac{V_{fest}}{V} = 1 - \frac{\pi \cdot d^2 \cdot l}{4 \cdot s_1 \cdot d \cdot l} = 1 - \frac{\pi}{4 \cdot a} \qquad \text{for } b > 1 \qquad (3.38)$$

$$\Psi = 1 - \frac{V_{fest}}{V} = 1 - \frac{\pi \cdot d^2 \cdot l}{4 \cdot s_1 \cdot s_2 \cdot l} = 1 - \frac{\pi}{4 \cdot a \cdot b} \qquad \text{for } b < 1 \qquad (3.39)$$

The velocity in the *Reynolds* number is:

$$c_\psi = c_0 / \Psi \qquad (3.40)$$

The *Reynolds* number:

$$Re_{\psi, L'} = \frac{c_\psi \cdot L'}{\nu} \qquad (3.41)$$

With this *Reynolds* number the *Nußelt* number of a single tube is determined. For the different arrangements of tubes two further geometry functions are defined to take into account the arrangement of the tubes. The first function f_A takes into account the *arrangement of tubes in the bundle* the second function f_n the number of tube rows.

$$f_A = 1 + \frac{0.7 \cdot (b/a - 0.3)}{\Psi^{1.5} \cdot (b/a + 0.7)^2} \quad \text{aligned arrangement} \tag{3.42}$$

$$f_A = 1 + \frac{2}{3 \cdot b} \qquad \text{staggered arrangement} \tag{3.43}$$

Compared to a single tube the heat transfer coefficients in the first row are larger, but they are smaller than in the following rows. With separation of flow and vortex generation the turbulence increases and so also the heat transfer coefficients. This so called *first row effect* must be considered additionally. Tube bundles are often calculated from tube row to tube row, for this the local heat transfer coefficient of every tube row is separately calculated with the corresponding flow velocity, material properties and the tube row correction factor. In Figure 3.10 on left side the local tube row correction function f_j for the j-th row is given. Equation (3.44) calculates the corrector function f_j and Equation (3.45) that for the total bundle with n tube rows. With Equation (3.46) the *Nußelt* number Nu_j of the j-th row and with Equation (3.47) that of the bundle Nu_{Bundle} can be determined.

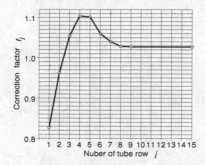 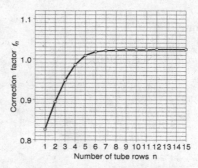

Figure 3.10: Correction function for the first row effect (left local, right integral)

$$f_j = \begin{vmatrix} 0.6475 + 0.2 \cdot j - 0.0215 \cdot j^2 & \text{if } j < 5 \\ 1 + 1/(j^2 + j) + 3 \cdot (2 \cdot j - 1)/(j^4 - 2 \cdot j^3 + j^2) & \text{if } j > 4 \end{vmatrix} \tag{3.44}$$

$$f_n = \begin{vmatrix} 0.74423 + 0.8 \cdot n - 0.006 \cdot n^2 & \text{if } n < 6 \\ 0.018 + \exp[0.0004 \cdot (n-6)) - 1] & \text{if } n > 5 \end{vmatrix} \tag{3.45}$$

$$Nu_j = \alpha \cdot L' / \lambda = Nu_{L'} \cdot f_A \cdot f_j \tag{3.46}$$

$$Nu_{bundle} = \alpha \cdot L' / \lambda = Nu_{L'} \cdot f_A \cdot f_n \tag{3.47}$$

EXAMPLE 3.6: Design of a nuclear reheater bundle

For a nuclear steam turbine unit a reheater bundle with U-tubes shall be designed.
The mass flow rate of the steam through the bundle is 300 kg/s and the inlet pressure
8 bar. The working steam shall be heated from 170.4 °C to 280 °C by heating steam
condensing at a temperature of 295 °C. The heat transfer coefficient in the tubes is
12 000 W/(m² K). The outside diameter of the tubes is 15 mm, wall thickness 1 mm and
thermal conductivity 26 W/(m K). To keep the pressure drop in an acceptable range,
the steam velocity to the bundle shall not exceed 6 m/s. The heating steam chamber
is a hemisphere welded to the circular tube sheet. The tubes shall fill a rectangular
space, i.e. the height of the bundle shall be approximately the same as its width. The
U-bends of tubes are not in contact with the steam i.e. they do not participate in the
heat transfer and therefore only the straight tube part between tube sheet and last
support plate exchange heat. The tubes are arranged in 60° triangles and have a
distance of 20 mm. The sketch shows the arrangement of the tube bundle and that of
the tubes. At the bundle inlet the steam density of steam is: 4.161 kg/m³.
The other steam properties at 225.2 °C are:

ρ = 3.581 kg/m³, λ = 0.038 W/(m K), ν = 4.76 · 10⁻⁶ m²/s, Pr = 0.99,
c_p = 2206 J/(kg K).

Find

Number and length of the tubes.

Solution

Schematic   See sketch.

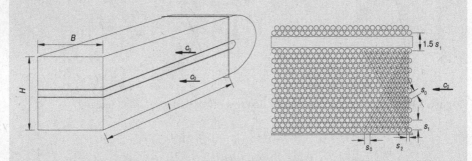

Assumptions

- Effects at the bundle periphery can be neglected.
- The mean heat transfer coefficients in- and outside of the tubes are constant.
- The temperature in the tubes is constant.

Analysis

To determine the number and length of the tubes, the number of tube rows n, the number of the tubes per row i and the heat transfer surface area A has to be determined. There are three unknowns, therefore three equations are required.

The condition that the height and width of the bundle must have the same dimension, defines the relationship between the number of tube rows n and number of tubes per row i and deliver the first equation.

$$B = n \cdot s_2 \qquad H = (i+1.5) \cdot s_1 \quad \text{from } B = H \text{ we receive: } n \cdot s_2 = (i+1.5) \cdot s_1$$

The given flow velocity at the bundle inlet determine the flow cross-sectional area at bundle inlet.

$$c_0 = \frac{\dot{m}}{H \cdot l \cdot \rho_0} = \frac{\dot{m}}{(i+1.5) \cdot s_1 \cdot l \cdot \rho_0}$$

The heat exchanger surface area can be calculated with the rate equation.

$$A = i \cdot n \cdot \pi \cdot d_a \cdot l = \frac{\dot{Q}}{k \cdot \Delta\vartheta_m}$$

With the energy balance equation the heat rate can be determined.

$$\dot{Q} = \dot{m} \cdot c_p \cdot (\vartheta_1'' - \vartheta_1') = 300 \cdot \text{kg/s} \cdot 2\,206 \cdot \text{J/(kg} \cdot \text{K)} \cdot (280 - 170.4) \cdot \text{K} = 72\,533 \text{ kW}$$

The log mean temperature difference is calculated with the given temperatures:

$$\Delta\vartheta_m = \frac{\vartheta_1'' - \vartheta_1'}{\ln\left(\dfrac{\vartheta_2 - \vartheta_1'}{\vartheta_2 - \vartheta_1''}\right)} = \frac{(280 - 170.4) \cdot \text{K}}{\ln\left(\dfrac{295 - 170.4}{295 - 280}\right)} = 51.77 \text{ K}$$

To determine the heat transfer coefficient first the influence of the bundle geometry has to be given. The dimensionless tube distances a and b are:

$$s_1 = \sqrt{3} \cdot s_0 = \sqrt{3} \cdot 20 \text{ mm} = 34.64 \text{ mm} \qquad a = s_1 / d_a = 34.64/15 = 2.309$$

$$s_2 = s_0 / 2 = 10 \text{ mm} \qquad b = s_2 / d_a = 10/15 = 0.67$$

As $b < 1$ the hollow volume ratio is calculated with Equation (3.39).

$$\Psi = 1 - \frac{\pi}{4 \cdot a \cdot b} = 1 - \frac{\pi}{4 \cdot a \cdot b} = 1 - \frac{\pi}{4 \cdot 2.309 \cdot 0.67} = 0.490$$

The flow length of the tube is: $L' = \pi \cdot d_a / 2 = 23.562$ mm

Due to the temperature rise, the density of steam in the bundle is smaller than at the inlet. This has to be considered in the determination of the velocities. The steam density is the one at the main steam temperature. With Equation (3.40) we receive:

$$c_\psi = \frac{c_0 \cdot \rho_0}{\rho \cdot \Psi} = \frac{6 \cdot 4.161}{3.581 \cdot 0.490} \cdot \frac{m}{s} = 14.23 \ \frac{m}{s}$$

Reynolds number with Equation (3.41):

$$Re_{L',\psi} = \frac{c_\psi \cdot L'}{\nu} = \frac{14.23 \cdot m \cdot 0.02356 \cdot m \cdot s}{4.76 \cdot 10^{-6} \cdot m^2 \cdot s} = 70\,448$$

The *Nußelt* number is determined with Equations (3.31) to (3.33), whereas it is assumed that the steam side heat transfer coefficient is rather small and therefore a wall temperature of 270 °C is near to reality.

$$Nu_{L',lam} = 0.664 \cdot \sqrt[3]{Pr} \cdot \sqrt{Re_{L'}} = 0.664 \cdot \sqrt[3]{0.99} \cdot \sqrt{70\,448} = 175.7$$

$$Nu_{L',turb} = \frac{0.037 \cdot Re_{L'}^{0.8} \cdot Pr}{1 + 2.443 \cdot Re_{L'}^{-0.1} \cdot (Pr^{2/3} - 1)} = \frac{0.037 \cdot 70\,448^{0.8} \cdot 0.99}{1 + 2.443 \cdot 70\,448^{-0.1} \cdot (0.99^{2/3} - 1)} = 278.3$$

$$Nu_{L'} = Nu_{L',0} + \sqrt{Nu_{L',lam}^2 + Nu_{L',turb}^2} \cdot (T / T_W)^{0.121} = 325.9$$

The function f_A is calculated with Equation (3.43).

$$f_A = 1 + \frac{2}{3 \cdot b} = 1 + \frac{2}{3 \cdot 0.67} = 2$$

For the function f_n it is first assumed that more than 15 tube rows would be required, resulting in $f_n = 1.03$. The *Nußelt* number of the bundle with Equation (3.47):

$$Nu_{bundle} = Nu_{L'} \cdot f_A \cdot f_n = 325.9 \cdot 2 \cdot 1.03 = 671.2$$

For the heat transfer coefficient outside the tubes we receive:

$$\alpha_a = \frac{Nu_{bundle} \cdot \lambda}{L'} = \frac{671.2 \cdot 0.038 \cdot W}{0.02356 \cdot m \cdot m \cdot K} = 1082.6 \ \frac{W}{m^2 \cdot K}$$

The overall heat transfer coefficient is:

$$k = \left(\frac{1}{\alpha_a} + \frac{d_a}{2 \cdot \lambda_R} \cdot \ln \frac{d_a}{d_i} + \frac{d_a}{d_i \cdot \alpha_i} \right)^{-1} = 942.4 \ \frac{W}{m^2 \cdot K}$$

To check the assumption made for the wall temperature, it is calculated now.

$$\vartheta_W = \vartheta_m + \Delta\vartheta_m \cdot k / \alpha_a = 225.2 \ °C + 51.77 \cdot K \cdot 942.4 / 1082.6 = 270.3 \ °C$$

A further correction is not necessary as the difference is less than 0.05 %. The required surface area is:

$$A = i \cdot n \cdot \pi \cdot d_a \cdot l = \frac{\dot{Q}}{k \cdot \Delta\vartheta_m} = \frac{72.533 \cdot 10^6 \cdot W \cdot m^2 \cdot K}{942.4 \cdot W \cdot 51.77 \cdot K} = 1486.7 \ m^2$$

From the equation for the steam velocity at the inlet of the bundle we receive:

$$(i+1.5) \cdot l = \frac{\dot{m}}{s_1 \cdot c_0 \cdot \rho_0} = \frac{300 \cdot kg \cdot s \cdot m^3}{0.03464 \cdot m \cdot 6 \cdot m \cdot 4.161 \cdot kg \cdot s} = 346.88 \ m$$

With the condition, that the height and width of the bundle have approximately the same dimension, the following equation was given:

$$(i+1.5) = n \cdot s_2 / s_1$$

The last two equations combined deliver:

$$n \cdot l = s_1 / s_2 \cdot 346.88 \ m = 0.03464 / 0.010 \cdot 346.89 \ m = 1201.63 \ m$$

With required surface area the number of tubes per rows can be determined. For the number of tubes we receive:

$$n = (i+1.5) \cdot s_1 / s_2 = \mathbf{96}$$

The tube length is: $l = 1201.67 \ m / n = \mathbf{12.498 \ m}$

Discussion

The heat transfer coefficient in the bundle is higher than on a single tube. This is due to the higher velocity and subsequently higher *Reynolds* number and due to the flow separation and vortex formation which are considered by the functions f_A and f_n. This example shows, that the design of a heat exchanger requires not only the knowledge of the heat transfer coefficient, but also knowledge of the mechanical design and pressure drop (the given flow velocity is a result of a pressure drop calculation).

3.2.5   Tube bundle with baffle plates

Often the tubes in bundles are not hit by the flow perpendicularly or at a given angle, due to so-called *baffle plates*. As shown in Figure 3.11, the flow in the bundle changes its direction, directed by the *baffle plates*. Areas with perpendicular and parallel flow exist and these two flow modes can be calculated for the perpendicular flow as shown in the chapter before and for the parallel flow as discussed in Chapter 3.2.13. this calculation is only a rough approach as close to the baffle plate ends neither parallel nor perpendicular flow exist. Furthermore at the baffle plates by-passes and leakages occur. These effects can be considered by correction functions, whose discussion would go beyond the scope of this book. Literature can be found e.g. in VDI Heat Atlas [3.4, 3.6].

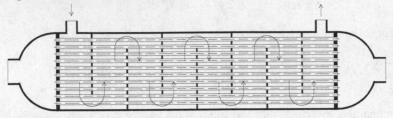

Figure 3.11: Tube and shell heat exchanger with baffle plates

3.3   Finned tubes

By installing fins, transfer areas of heat exchangers gain *extended surfaces*. This is a fairly cost-efficient solution, as for the extended surface area no pressure piping is required. Fins are normally always installed on the side of lower heat transfer coefficients. The benefit of the fins is as greater, the lower the heat transfer coefficients are.

In the following calculations a perfect thermal contact between the fins and the base wall is assumed. In any case, this is an absolute requirement for manufacturing finned surfaces. The calculation procedures discussed here are not generally exact, as in reality no analytical calculation methods have been found yet to predict the heat transfer coefficients for all types of finned surfaces. For exact calculations either numerical simulations, tests or reports of tests, performed in similar conditions, are required.

For plain plates with fins of constant cross-sections the overall heat transfer coefficients can be determined as given in Chapter 3.2.2.

Figure 3.12 shows the arrangement and fin shapes of typical finned heat transfer surface areas.

In this chapter only *finned tubes* will be discussed in detail. The heat transfer coefficients are related to the outer surface area A of the non-finned tube. Therefore, the heat rate of a finned tube is given as:

$$\dot{Q} = k \cdot A \cdot \Delta\vartheta_m = k \cdot \pi \cdot d_a \cdot l \cdot \Delta\vartheta_m \tag{3.48}$$

The overall heat transfer coefficient related to the outer surface area of the non-finned tube A is determined as follows: The heat transfer coefficient at the fin surface A_{Ri} and at the surface A_0 of the tubes between the fins is α_a. The related temperature is that of the outer surface of the tube. The changed temperatures on the fin surface are considered by the fin effectiveness η_{Ri}. The *fin surface area* A_{Ri} is that of the fin walls whereas the surface area of the fin tip is neglected. This results in the following overall heat transfer coefficient:

$$\frac{1}{k} = \frac{A}{A_0 + A_{Ri} \cdot \eta_{Ri}} \cdot \frac{1}{\alpha_a} + \frac{d_a}{2 \cdot \lambda_R} \cdot \ln\frac{d_a}{d_i} + \frac{d_a}{d_i} \cdot \frac{1}{\alpha_i} \tag{3.49}$$

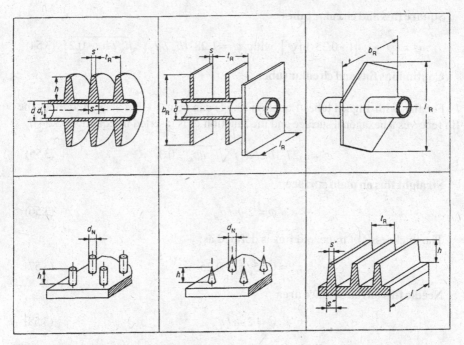

Figure 3.12: Typical finned surfaces [3.4]

The fin efficiency shown in Chapter 2.1.6.4 is only valid for fins with a constant cross-section. Due to the not constant cross-section of finned tubes, a new correlation for the *fin efficiency* is needed. They will be given below.

$$\eta_{Ri} = \frac{\tanh X}{X} \tag{3.50}$$

The dimensionless value X is calculated as:

$$X = \varphi \cdot \frac{d_a}{2} \cdot \sqrt{\frac{2 \cdot \alpha_a}{\lambda \cdot s_{Ri}}} \tag{3.51}$$

The correction function for different geometries is φ. For conical fins the thickness of fin s is the average value of the thickness at the fin foot s'' and fin tip s'.

$$s_{Ri} = (s'' + s')/2 \tag{3.52}$$

The following correction functions φ are given below:

Annular fins:

$$\varphi = (D/d_a - 1) \cdot \left[1 + 0.35 \cdot \ln(D/d_a)\right] \tag{3.53}$$

Square fins and circular tubes

$$\varphi = (\varphi' - 1) \cdot \left[1 + 0.35 \cdot \ln \varphi'\right] \text{ with } \varphi' = 1.28 \cdot (b_R/d_a) \cdot \sqrt{l_R/b_R - 0.2} \tag{3.54}$$

Continuous fins and circular tubes

For fins with aligned tubes Equation (3.54) can be used. For staggered tubes the fin receives a hexagonal surface and the function φ' is inserted in Equation (3.54):

$$\varphi' = 1.27 \cdot (b_R/d_a) \cdot \sqrt{l_R/b_R - 0.3} \tag{3.55}$$

Straight fins on plain surface

$$\varphi = 2 \cdot h/d_a \tag{3.56}$$

The thickness s of trapezoid fins is defined as:

$$s_{Ri} = 0.75 \cdot s'' + 0.25 \cdot s' \tag{3.57}$$

Needle fins on flat surface area

$$\varphi = 2 \cdot h/d_a \tag{3.58}$$

The thickness s of *needle fins* is defined as:

$$s_{Ri} = d_N/2 \text{ at blunt}, s_{Ri} = 1.125 \cdot d_N \text{ at sharp fins} \tag{3.59}$$

3.3.1   Annular fins

The determination of the surface areas of annular fins with constant fin thickness is discussed below. For rectangular and continuous fins the procedure is similar. Figure 3.13 shows a circular tube with *annular fins* of constant thickness.

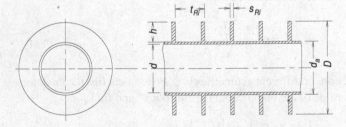

Figure 3.13: Annular fins of constant thickness

The surface area A of the non-finned tube is:

$$A = \pi \cdot d_a \cdot l \tag{3.60}$$

The surface area A_0 of the tube between the fins is:

$$A_0 = \pi \cdot d_a \cdot l \cdot (1 - s_{Ri} / t_{Ri}) \tag{3.61}$$

Surface area of the fins A_{Ri}:

$$A_{Ri} = 2 \cdot \frac{\pi}{4} \cdot (D^2 - d_a^2) \cdot \frac{l}{t_{Ri}} \tag{3.62}$$

The ratio of the fin surface area to that of the unfinned tube is:

$$\frac{A_{Ri}}{A} = \left[(D / d_a)^2 - 1 \right] \cdot \frac{d_a}{2 \cdot t_{Ri}} = \frac{2 \cdot h \cdot (d_a + h)}{t_{Ri} \cdot d_a} = \frac{2 \cdot h}{t_R} \cdot (1 + h / d_a) \tag{3.63}$$

Heat transfer coefficients for tube bundles as shown in the previous chapter can not be applied for finned tubes. Therefore here a correlation for tube bundles with finned tubes is given, which reproduce a large number of test values with an accuracy of 10 to 25 % [3.6].

$$Nu_{d_a} = C \cdot Re_{d_a}^{0.6} \cdot \left[(A_{Ri} + A_0) / A \right]^{-0.15} \cdot Pr^{1/3} \cdot f_4 \cdot f_n \tag{3.64}$$

The constant C for aligned tube arrangement is $C = 0.22$ and for shifted $C = 0.38$. The characteristic length for *Nußelt* and *Reynolds* number is the outside diameter of the tube.

The *Reynolds* number is determined with the velocity in the narrowest gap between the tubes.

The narrowest gap is dependent on the arrangement of the tubes as demonstrated in Figure 3.14. At the determination of the narrowest gap, the obstruction by the fins has to be considered.

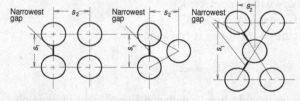

Figure 3.14: Determination of the narrowest gap

For the two left-hand arrangements in Figure 3.14 the velocity in the narrowest gap results as:

$$c_e = c_0 \cdot \left[(1 - \frac{1}{a}) - \frac{2 \cdot s_{Ri} \cdot h}{s_1 \cdot t_{Ri}} \right]^{-1} \tag{3.65}$$

For the arrangement on the right-hand side we receive:

$$c_e = c_0 \cdot \left[\sqrt{1 + (2 \cdot b / a)^2} - \frac{2}{a} - \frac{4 \cdot s_{Ri} \cdot h}{s_1 \cdot t_{Ri}} \right]^{-1} \tag{3.67}$$

More exact correlations for low finned tubes are proposed by *Briggs* and *Young* [3.7], based on a large number of tests.

EXAMPLE 3.7: Reheater bundle with finned tubes

A reheater bundle with finned tubes has to be designed for the same thermal and geometrical conditions as given in Example 3.6. The fins have an outside diameter of 5/8". The height of the fins is 1.27 mm, the thickness 0.3 mm and their distance 1 mm. The thermal conductivity of tube and fin material is 27 W/(m K), the tube wall thickness 1 mm, the distance between the tubes 13/16".

Find

The number and the length of the tubes.

Solution

Schematic   See sketch.

Assumptions

- The effect at bundle periphery can be neglected.
- All mean heat transfer coefficients are constant.
- The condensation temperature in the tubes is constant.

Analysis

The procedure is the same as for the tube bundle with non-finned tubes in Example 3.6. First the dimension given in U.S. unit are converted to SI units.

$$D = 5/8" \cdot 25.4 \cdot mm = 15.875 \text{ mm} \quad s_0 = 13/16" \cdot 25.4 \cdot mm = 20.6375 \text{ mm}$$

The required geometrical values are:

$$d_a = D - 2 \cdot s = (15.875 - 2 \cdot 1.27) \cdot \text{mm} = 13.335 \ \text{mm} \quad d_i = 11.335 \ \text{mm}$$

$$s_1 = s_0 \cdot \sqrt{3} = 35.7452 \ \text{mm} \qquad s_2 = s_0 / 2 = 10.31875 \ \text{mm}$$

$$a = s_1 / d_a = 2.681 \qquad b = s_2 / d_a = 0.77381$$

The velocity is determined with the steam density in the bundle using Equation (3.66).

$$c_e = c_0 \cdot \frac{\rho_0}{\rho} \cdot \left[\sqrt{1 + (2 \cdot b / a)^2} - \frac{2}{a} - \frac{4 \cdot s_{Ri} \cdot h}{s_1 \cdot t_{Ri}} \right]^{-1} =$$

$$= 6 \cdot \frac{\text{m}}{\text{s}} \cdot \frac{4.161}{3.581} \cdot \left[\sqrt{1 + (2 \cdot 0.7738 / 2.681)^2} - \frac{2}{2.681} - \frac{4 \cdot 0.3 \cdot 1,27}{35.7452 \cdot 1} \right]^{-1} = 19.05 \ \frac{\text{m}}{\text{s}}$$

The *Reynolds* number is calculated with this velocity and the outer tube diameter.

$$Re_{d_a} = \frac{c_e \cdot d_a}{\nu} = \frac{19.05 \cdot 0.013335}{4.76 \cdot 10^{-6}} = 53371$$

Before the *Nußelt* number can be calculated with Equation (3.64), first the surface areas and correction functions have to be calculated. For f_4 and f_n the values used in Example 3.6 are inserted.

$$\frac{A_{Ri}}{A} = \frac{2 \cdot h}{t_R} \cdot (1 + h / d_a) = \frac{2 \cdot 1.27}{1} \cdot (1 + 1.27 / 13.335) = 2.7819$$

$$\frac{A_0}{A} = 1 - s / t_R = 0.7$$

$$Nu_{d_a} = 0.38 \cdot Re_{d_a}^{0.6} \cdot [(A_{Ri} + A_0) / A]^{-0.15} \cdot Pr^{1/3} \cdot f_4 \cdot f_n =$$

$$= 0.38 \cdot 53371^{0.6} \cdot 3.4819^{-0.15} \cdot 0.99^{1/3} \cdot 0.99 \cdot 1.03 = 219.7$$

The heat transfer coefficient outside the tubes is:

$$\alpha_a = Nu_{d_a} \cdot \lambda / d_a = 626.2 \ \text{W/(m}^2 \cdot \text{K)}$$

For the determination of the overall heat transfer coefficient the fin efficiency has to be calculated with Equations (3.50), (3.51) and (3.53).

$$\varphi = (\frac{D}{d_a} - 1) \cdot \left[1 + 0.35 \cdot \ln \left(\frac{D}{d_a} \right) \right] = \left(\frac{15.875}{13.335} - 1 \right) \cdot \left[1 + 0.35 \cdot \ln \left(\frac{15.875}{13.335} \right) \right] = 0.2021$$

$$X = \varphi \cdot \frac{d_a}{2} \cdot \sqrt{\frac{2 \cdot \alpha_a}{\lambda \cdot s_{Ri}}} = 0.2021 \cdot \frac{0.013335 \cdot \text{m}}{2} \cdot \sqrt{\frac{2 \cdot 626.2 \cdot \text{W} \cdot \text{m} \cdot \text{K}}{27 \cdot \text{W} \cdot 0.0003 \cdot \text{m} \cdot \text{m}^2 \cdot \text{K}}} = 0.540$$

$$\eta_{Ri} = \frac{\tanh X}{X} = \frac{\tanh 0.540}{0.540} = 0.913$$

$$k = \left(\frac{1}{0.7 + 2.7819 \cdot 0.913} \cdot \frac{1}{626.2} + \frac{0.013335}{2 \cdot 27} \cdot \ln \frac{13.335}{11.335} + \frac{13.335}{11.335} \cdot \frac{1}{12\,000} \right)^{-1} =$$

$$= 1580.7 \ \frac{\text{W}}{\text{m}^2 \cdot \text{K}}$$

The required heat exchanging surface area is:

$$A = i \cdot n \cdot \pi \cdot d_a \cdot l = \frac{\dot{Q}}{k \cdot \Delta \vartheta_m} = \frac{72.533 \cdot 10^6 \cdot \text{W} \cdot \text{m}^2 \cdot \text{K}}{1580.7 \cdot \text{W} \cdot 51.77 \cdot \text{K}} = 886.37 \ \text{m}^2$$

With the given inlet velocity we receive:

$$(i+1.5) \cdot l = \frac{\dot{m}}{s_1 \cdot c_0 \cdot \rho_0} = \frac{300 \cdot \text{kg} \cdot \text{s} \cdot \text{m}^3}{0.0357452 \cdot \text{m} \cdot 6 \cdot \text{m} \cdot 4.161 \cdot \text{kg} \cdot \text{s}} = 336.17 \ \text{m}$$

From the ratio of the bundle height to width results:

$$\cdot \quad (i+1.5) = n \cdot s_2 / s_1$$

The combination of both equations deliver:

$$n \cdot l = 336.17 \ \text{m} \cdot s_1 / s_2 = 336.17 \ \text{m} \cdot 35.7452 / 10.31875 = 1164.52 \ \text{m}$$

From the surface area we receive the number of tubes per row.

$$i = \frac{A}{n \cdot l \cdot \pi \cdot d_a} = \frac{886.37 \ \text{m}^2}{1164.52 \cdot \text{m} \cdot \pi \cdot 0.013335 \cdot \text{m}} = 18$$

The number of tubes results as: $n = (i+1.5) \cdot s_1 / s_2 = \mathbf{68}$

$$l = 1164.52 \ \text{m} / n = \mathbf{17.125 \ m}$$

Discussion

At a first glance it seems that with the finned tubes the surface area is 67 % smaller than with the non-finned tubes. The finned tube surface area was calculated with the outside diameter of the tubes. The total surface area with fins is 3.48 times larger and is 3080 m^2. The most important fact is, the bundle is smaller than the one with non-finned tubes. Instead of 1 262 only 612 U-tubes are required. The height and width of the bundle is reduced from 1 m to 0.7 m. The tube lengths are indeed longer, they increase from 12.5 m to 17.125 m. Manufacturing costs for longer bundles are lower, as fewer tubes are needed. This reduces the effort for drilling the holes in the thick tube sheet, welding and bending of the tubes.

It appears paradoxical, that the overall heat transfer coefficient is larger than that of the steam. The heat transfer occurs on the tube surface between the fins and on the fin surface. The overall heat transfer coefficient is related to the non-finned surface of the tubes and is therefore larger.

For this type of so called "low finned tubes" a correlation reported by *Briggs* and *Young* [3.8] may provide more exact results.

4   Free convection

Contrary to forced convection, in *free convection* the flow is generated only by the temperature difference between wall and fluid. When a static fluid contacts a surface (wall) which has a different temperature to the fluid, temperature differences will be created in the fluid and subsequently the fluid density will change. Due to gravity, fluid layers with higher temperature ascend, such with lower temperature descend. Thermal and hydraulic boundary layers are generated by the temperature and subsequent density difference.

The analysis of a huge number of test and flow models delivered empiric equations for the *Nußelt* number as a function of the *Grashof* and *Prandtl* number.

$$Nu_l = \alpha \cdot L / \lambda = f(Gr, \text{Pr})$$

The *Grashof number* is the ratio between the buoyancy and frictional forces. It is describing the similar correlations as the *Reynolds* number for forced convection. It is defined as:

$$Gr = \frac{g \cdot L^3 \cdot (\rho_W - \rho_0)}{\rho_0 \cdot v^2} \tag{4.1}$$

Herein L is the characteristic length, β the thermal expansion coefficient and v the kinematic viscosity. The index W is for the state at the wall, 0 that of the static fluid.

For ideal gases the thermal expansion coefficient is only a function of the absolute temperature of the gas.

$$\beta = 1/T_0 \tag{4.2}$$

For very small values of $\beta \cdot (\vartheta_W - \vartheta_0) \ll 1$ the *Grashof* number can be given as a function of the density difference.

$$\frac{(\rho_W - \rho_0)}{\rho_0} = \beta \cdot (\vartheta_W - \vartheta_0) \tag{4.3}$$

$$Gr = \frac{g \cdot l^3 \cdot \beta \cdot (\vartheta_W - \vartheta_0)}{v^2} \tag{4.4}$$

The mean temperature $(\vartheta_W - \vartheta_0)/2$ is used to determine the material properties λ, v and Pr.

The characteristic length l in the *Grashof* and *Nußelt* number is:

119

$$L = A/U_{proj} \tag{4.5}$$

The heat transfer surface area is A and the projected circumference in direction of the flow U_{proj}. For simplifying the equations, the *Rayleigh number* will be established as a further similarity number. It is the product of *Grashof* and *Prandtl* number.

$$Ra = Gr \cdot Pr \tag{4.6}$$

Equations for the *Nußelt* numbers published in literature [4.1] are valid for constant wall temperatures. However, in practice the deviation to a calculation with the mean wall temperature is negligible.

4.1   Free convection at plain vertical walls

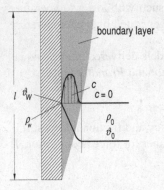

Figure 4.1: Free convection at a plain vertical wall

At a heated plain vertical wall (Figure 4.1) of the height l, the density of the fluid in the boundary layer close to the wall is lower, therefore it is subject to buoyancy and an ascended flow develops. At a wall with cooling a descendent flow occurs. In a stationary state the buoyant and frictional forces are balanced. The flow is first laminar and after a certain length turbulent. In the boundary layer the temperature changes with the wall distance and with it the density of the fluid. Therefore, the buoyancy in the fluid layers are different. Because of this, generally applicable analytical solutions of the flow and energy equations have not yet been found. The following empiric correlation delivers mean *Nußelt* numbers.

The characteristic length of the vertical wall is:

$$L = \frac{b \cdot l}{b} = l$$

For plain vertical walls the following empirical correlation is proposed:

$$Nu_l = \left\{ 0.825 + 0.387 \cdot Ra^{1/6} \cdot f_1(Pr) \right\}^2 \tag{4.7}$$

$$f_1(Pr) = \left(1 + 0.671 \cdot Pr^{-9/16}\right)^{-8/27} \tag{4.8}$$

The validity range of the above equations is:

$$0.001 < Pr < \infty$$
$$0.1 < Ra < 10^{12}$$

Equation (4.7) is valid for laminar and turbulent flow.

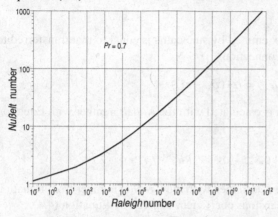

Figure 4.2: *Nußelt* vs. *Rayleigh* number for $Pr = 0.7$

Figure 4.2 shows the *Nußelt* versus *Rayleigh* number. At *Rayleigh* numbers higher than 10^6 the *Nußelt* number is proportional to the third power of the temperature difference.

EXAMPLE 4.1: Warming of a vertical wall

The 3 m high wall of a house is heated by a heat flux of 100 W/m² originating from sunlight. The temperature of the ambient air is 0 °C. The material properties of the air are: $\lambda = 0.0245$ W/(m K), $\nu = 14 \cdot 10^{-6}$ m²/s, $Pr = 0.711$.

Find

The temperature of the wall.

Solution

Schematic   See sketch.

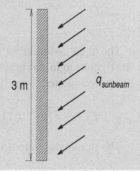

Assumptions

- The warming of the wall from the inside is neglected.
- Radiation emitted from the wall is neglected.
- Only the steady state, i.e. the state at which the wall has reached its final temperature, shall be discussed here.

Analysis

In a steady state the heat flux emitted by sunbeams is equal to that transferred to the environmental air by free connection.

$$\dot{q}_{sun} = \alpha \cdot (\vartheta_w - \vartheta_0)$$

The heat transfer coefficient is calculated with the *Nußelt* number with Equation (4.7).

$$\alpha = Nu_l \cdot \lambda / l = \left\{ 0.825 + 0.387 \cdot Ra^{1/6} \cdot f_1(Pr) \right\}^2 \cdot \lambda / l$$

The *Rayleigh* number can be determined with the *Grashof* number from Equation (4.3), whereas the thermal expansions coefficient is given by Equation (4.4).

$$Gr = \frac{g \cdot l^3 \cdot (\vartheta_w - \vartheta_0)}{T_0 \cdot v^2}$$

The *Grashof* number and subsequently the heat transfer coefficient are functions of the wall temperature. For the determination of the wall temperature two methods are possible: in the first one a wall temperature is estimated to calculate the heat transfer coefficient and with it the wall temperature. The calculation procedure has to be repeated until the required accuracy is reached. In the second method the temperature difference is inserted as the ratio of the heat flux and heat transfer coefficient in Equation (4.4) and the heat transfer can be calculated directly. The second procedure delivers:

$$\alpha = \left\{ 0.825 + 0.387 \cdot \left(\frac{g \cdot l^3 \cdot \dot{q}}{\alpha \cdot T_0 \cdot v^2} \right)^{1/6} \cdot Pr^{1/6} \cdot f_1(Pr) \right\}^2 \cdot \frac{\lambda}{l}$$

The exact solution must be calculated with an equation solver or by iteration. If the right term in the brackets is much higher than 0.825, the heat transfer coefficient can be calculated directly. The function $f_1(Pr)$ is determined with Equation (4.7).

$$f_1(Pr) = \left(1 + 0.671 \cdot Pr^{-9/16}\right)^{-8/27} = \left(1 + 0.671 \cdot 0.711^{-9/16}\right)^{-8/27} = 0.8384$$

The numerical values inserted in *Mathcad* deliver:

$$\alpha = \left\{ 0.825 + 0.387 \cdot \left(\frac{g \cdot l^3 \cdot q_{sun}}{\alpha \cdot T_0 \cdot v^2} \right)^{1/6} \cdot Pr^{1/6} \cdot f(Pr) \right\}^2 \cdot \frac{\lambda}{1} = 4.1 \ \frac{W}{m^2 \cdot K}$$

The wall temperature we receive as:

$$\vartheta_W = \vartheta_0 + \dot{q} / \alpha = \mathbf{24.4 \ °C}$$

Discussion

Free convection is often generated by external influences in this example by sunlight. If the wall temperature is not a given value, an iteration is required for its determination. In the case of given heat flux or heat flow rate the temperature difference can be replaced by the ratio of heat flux and heat transfer coefficient.

EXAMPLE 4.2: Radiator

In a room with radiators of 1.2 m length, 0.45 m height and 0.02 m width a heating power of 3 kW shall be established. The mean heater wall temperature is 48 °C, the room temperature 22 °C. Material properties of air:
 $\lambda = 0.0268$ W/(m K), $v = 16.05 \cdot 10^{-6}$ m²/s, $Pr = 0.711$.

Find

The number of radiators required for 3 kW heat rate.

Solution

Schematic   See sketch.

Assumptions

- The top and bottom wall with 20 mm width will not be considered.
- Radiation effects will not be considered.
- The wall temperature of the radiator is assumed as constant.

Analysis

The heat transfer surface area is:

$$A = 2 \cdot (H \cdot L + H \cdot B) = 2 \cdot (0.45 \cdot 1.2 + 0.45 \cdot 0.02) \ m^2 = 1.098 \ m^2$$

The *Rayleigh* number is calculated with the *Grashof* number from Equation (4.3). *Rayleigh* number will be determined with Equation (4.4).

$$Ra = \frac{g \cdot H^3 \cdot (\vartheta_W - \vartheta_0)}{T_0 \cdot v^2} \cdot Pr = \frac{9.81 \cdot m \cdot 0.45^3 \cdot m^3 \cdot (48-22) \cdot K \cdot s^2}{295.15 \cdot K \cdot 16.05^2 \cdot 10^{-12} \cdot m^4 \cdot s^2} \cdot 0.711 = 2.173 \cdot 10^8$$

The function $f_1(Pr)$ can be calculated with Equation (4.8).

$$f_1(Pr) = \left(1 + 0.671 \cdot Pr^{-9/16}\right)^{-8/27} = \left(1 + 0.671 \cdot 0.711^{-9/16}\right)^{-8/27} = 0.8384$$

$$Nu_H = \left\{0.825 + 0.387 \cdot Ra^{1/6} \cdot f_1(Pr)\right\}^2 =$$

$$= \left\{0.825 + 0.387 \cdot (2.173 \cdot 10^8)^{1/6} \cdot 0.8384\right\}^2 = 77.10$$

$$\alpha = Nu_L \cdot \frac{\lambda}{H} = 77.10 \cdot \frac{0.0268 \cdot W}{0.45 \cdot m \cdot m \cdot K} = 4.59 \ \frac{W}{m^2 \cdot K}$$

The heat flow rate delivered by each heating element is:

$$\dot{Q}_1 = \alpha \cdot A \cdot (\vartheta_W - \vartheta_0) = 4.59 \cdot W/(m^2 \cdot K) \cdot 1.098 \cdot m^2 \cdot (48-22) \cdot K = \mathbf{131.1\ W}$$

To produce 3 kW heat rate **23** heating elements are required.

Discussion

With known wall temperature the determination of the heat transfer coefficient and heat flux is fairly simple. The calculated value is not realistic however, because the heat transferred by radiation has not been considered. As will be shown in the chapter "Radiation" that almost 50 % of the heat rate is additionally transferred by radiation. At higher wall temperatures of the heater the portion of radiation increases. This type of heating elements are therefore called radiators.

EXAMPLE 4.3: Wall temperature of a room

In a room with an in- and outside wall with a height of 2.8 m, the room temperature is 22 °C, and the outer temperature 0 °C. The wall heat transfer coefficient has the value of 0.3 W/(m² K). Material properties of air:

inside: $\lambda = 0.0257$ W/(m K), $v = 15.11 \cdot 10^{-6}$ m²/s, $Pr = 0.713$
outside: $\lambda = 0.0243$ W/(m K), $v = 13.30 \cdot 10^{-6}$ m²/s, $Pr = 0.711$

Find

The temperature inside and outside on the wall.

Solution

Schematic   See sketch.

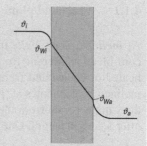

Assumptions

- Radiation effects are neglected.
- The wall temperatures are assumed to be constant.

Analysis

In this case first the wall temperatures has to be estimated, the heat transfer coefficients calculated and wall temperatures determined. With received wall temperatures the whole calculation has to be repeated until the required accuracy is reached. The results will be presented tabularly. The following equations were applied:

$$Ra = \frac{g \cdot l^3 \cdot (\vartheta_W - \vartheta_0)}{T_0 \cdot \nu^2} \cdot Pr$$

$$f_1(Pr) = \left(1 + 0.671 \cdot Pr^{-9/16}\right)^{-8/27}$$

$$Nu_l = \left\{ 0.825 + 0.387 \cdot Ra^{1/6} \cdot f_1(Pr)\right\}^2 \qquad \alpha = Nu_L \cdot \frac{\lambda}{L}$$

$$k = \left(\frac{1}{\alpha_a} + \frac{1}{\alpha_W} + \frac{1}{\alpha_i}\right) \quad \vartheta_{W_i} = \vartheta_i - (\vartheta_i - \vartheta_a) \cdot k / \alpha_i \quad \vartheta_{W_a} = \vartheta_a + (\vartheta_i - \vartheta_a) \cdot k / \alpha_a$$

The values of function $f_1(Pr)$ are in the room 0.8386 and outside 0.8384.

ϑ_{Wi} °C	ϑ_{Wa} °C	Ra_i $\cdot 10^{-9}$	Ra_a $\cdot 10^{-9}$	α_i	α_a W/(m² K)	k	ϑ_{Wi} °C	ϑ_{Wa} °C
20.00	2.00	5.880	4.909	1.960	1.750	0.227	19.46	2.85
19.46	2.85	7.467	6.995	2.113	1.956	0.232	19.59	2.60
19.59	2.60	7.085	6.381	2.078	1.901	0.230	19.56	2.67
19.56	2.67	7.173	6.553	2.086	1.917	0.231	19.57	2.65
19.57	2.65	7.144	6.504	2.084	1.912	0.231	**19.57**	**2.65**

Discussion

The magnitude of the heat transfer coefficients is determined by the temperature differences. Therefore the wall temperatures must be determined by iteration. The calculation of wall temperatures is important because they are relevant for convertibility and determine dew formation in the room, which should be strictly avoided.

4.1.1   Inclined plane surfaces

At *inclined plane walls* the flows initiated by the temperature difference can be com-pletely different from that at vertical walls. It depends whether surface is heated or cooled and from which side (upper or lower) the heat is transferred. At inclined heated plates at the lower side a stable boundary layer is generated which does not separate from the wall, from the upper side the boundary layer separates after a certain length. One has to distinguish between the following cases:

1. Heated surface with heat release from lower side: no boundary layer separation
2. Cooled surface with heat release from upper side: no boundary layer separation
3. Heated surface with heat release from upper side: boundary layer separation possible
4. Cooled surface with heat release from lower side: boundary layer separation possible.

If no boundary layer separation occurs (1. and 2.) Equation (4.7) can be used but the *Rayleigh* number must be multiplied by $\cos\alpha$, where α is the inclination angle to the vertical.

$$Ra_\alpha = Ra \cdot \cos\alpha \qquad\qquad (4.9)$$

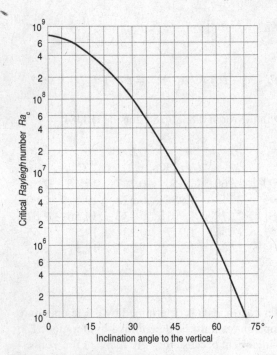

Figure 4.3: Critical *Rayleigh* number

In case of possible boundary layer separation (3. and 4.) the critical *Rayleigh* number Ra_c (Figure 4.3) define, which equation has to be used. For *Rayleigh* numbers lower than Ra_c, Equation (4.7) applies and for higher *Rayleigh* numbers the separation of the boundary layer has to be considered. For the separated boundary layer the following *Nußelt* number is prosed:

$$Nu_l = 0.56 \cdot (Ra_c \cdot \cos\alpha)^{1/4} + 0.13 \cdot (Ra^{1/3} - Ra_c^{1/3}) \qquad (4.10)$$

EXAMPLE 4.4: Solar collector

A solar collector panel with an angle of 45° to the vertical is installed on a roof. It has a length of 2 m and width of 1 m. The temperature of the collector is 30 °C, that of the surrounding air 10 °C. Material properties of air.
$\lambda = 0.0257$ W/(m K), $v = 15.11 \cdot 10^{-6}$ m²/s, $Pr = 0.713$.

Find

The heat losses on the upper side of the collector.

Solution

Schematic   See sketch.

Assumptions

- Radiation effects are not considered.
- The wall temperature of the collector is constant.

Analysis

The characteristic length of the panel: $L = 2$ m

$$Ra = \frac{g \cdot L^3 \cdot (\vartheta_w^2 - \vartheta_0^2) \cdot \cos\alpha}{T_0 \cdot v^2} \cdot Pr = \frac{9.81 \cdot \text{m} \cdot 2^3 \cdot \text{m}^3 \cdot (30-10) \cdot \text{K} \cdot \text{s}^2 \cdot 0.707}{283.15 \cdot \text{K} \cdot 15.11^2 \cdot 10^{-12} \cdot \text{m}^4 \cdot \text{s}^2} \cdot 0.713 = 1.22 \cdot 10^{10}$$

From Figure 4.3 at 45° the critical *Rayleigh* number is $Ra_c = 1.2 \cdot 10^7$. It is higher then the *Rayleigh* number of this case. The boundary layer separates, therefore the *Nußelt* number has to be calculated with Equation (4.10).

$$Nu_l = 0.56 \cdot (Ra_c \cdot \cos\alpha)^{1/4} + 0.13 \cdot (Ra^{1/3} - Ra_c^{1/3}) =$$
$$= 0.56 \cdot (1.2 \cdot 10^7 \cdot 0.707)^{1/4} + 0.13 \cdot [(1.22 \cdot 10^{10})^{1/3} - (1.2 \cdot 10^7)^{1/3}] = 299.3$$

$$\alpha = \frac{Nu_l \cdot \lambda}{l} = \frac{299.3 \cdot 0.0257 \cdot W}{2 \cdot m \cdot m \cdot K} = 3.84 \ \frac{W}{m^2 \cdot K}$$

The heat flow rate transferred from the upper side of the panel is:

$$\dot{Q} = \alpha \cdot A \cdot (\vartheta_w - \vartheta_0) = 4.384 \cdot W \cdot m^{-2} \cdot K^{-1} \cdot 2 \cdot m^2 \cdot (30 - 10) \cdot K = \mathbf{153} \ \mathbf{W}$$

Discussion

Without taking the flow separation into account, the heat transfer coefficient would be 3.28 W/(m² K), i.e. smaller. On a vertical plate it would be with 3.45 W/ (m² K), i.e. also smaller. The flow separation increases the heat transfer.

4.2   Horizontal plane surfaces

For heated horizontal plates transferring heat upwards, cooled surfaces receiving heat on their lower side the following equations were found:

$$Nu_l = 0.766 \cdot [Ra \cdot f_2(Pr)]^{1/5} \qquad \text{for} \qquad Ra \cdot f_2(Pr) \le 7 \cdot 10^4$$
$$Nu_l = 0.15 \cdot [Ra \cdot f_2(Pr)]^{1/3} \qquad \text{for} \qquad Ra \cdot f_2(Pr) > 7 \cdot 10^4 \qquad (4.11)$$

$$f_2(Pr) = \left(1 + 0.536 \cdot Pr^{-11/20}\right)^{-20/11} \qquad (4.12)$$

These correlations are valid for surfaces areas which are a part of an infinitely large surface area, i.e. the boundary layer is not disturbed by sideway restrictions.

The characteristic length l is given by Equation (4.5). For a square surface with the dimensions a and b, it is $l = a \cdot b/2 \ (a + b)$ and for a circular surface $l = d/4$.

Surface areas with side restrictions, e.g. under-floor heating, the above equations are not valid. Because the free convection on the side walls influence the heat transfer of the horizontal floor.

4.3   Free convection on contoured surface areas

Free convection will occur an all bodies with a temperature difference to the static surroundings without any restriction. The geometry influences the magnitude of the heat transfer. Here the *Nußelt* number will be given for a horizontal cylinder and for a sphere.

4.3.1   Horizontal cylinder

The *Nußelt* and *Rayleigh* number are determined with the same characteristic length as for forced convection with $L' = \pi \cdot d/2$. For the horizontal cylinder the following equations apply:

$$Nu_{L'} = \left[0.752 + 0.387 \cdot Ra_{L'}^{1/6} \cdot f_3(Pr) \right]^2 \qquad (4.13)$$

$$f_3(Pr) = (1 + 0.721 \cdot Pr^{-9/16})^{-8/27} \qquad (4.14)$$

EXAMPLE 4.5: Insulation of a steam pipe

A pipe with 100 mm outer diameter contains hot flowing steam with a temperature of 400 °C. According to the rules of the employers mutual insurance association at an air temperature of 30 °C. the surface temperature may not exceed 40 °C. It can be assumed that the surface of the steel pipe has the same temperature as the steam. The thermal conductivity of the insulation is 0.03 W/(m K). Material properties of the air:
$\lambda = 0.0265$ W/(m K), $v = 16.5 \cdot 10^{-6}$ m²/s, $Pr = 0.711$

Find

The diameter of the insulation.

Solution

Schematic   See sketch.

Assumptions

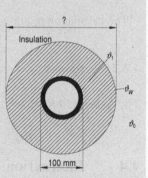

- The tube wall and the steam have the same temperature.
- The temperature on the outer surface of the insulation is constant.

Analysis

The temperature can be calculated with Equation (2.29).

$$\vartheta_W = \vartheta_0 + (\vartheta_1 - \vartheta_0) \cdot k / \alpha$$

The heat transfer coefficient α is determined with Equation (4.14), the overall heat transfer coefficient related to the outer surface area with Equation (2.27):

$$k = \left(\frac{1}{\alpha} + \frac{D}{2 \cdot \lambda_I} \cdot \ln \frac{D}{d} \right)^{-1}$$

$$\alpha = \left[0.752 + 0.387 \cdot Ra_{L'}^{1/6} \cdot f_3(Pr) \right]^2 \cdot \frac{\lambda}{L'} =$$

$$= \left[0.752 + 0.387 \cdot \left(\frac{g \cdot \pi^3 \cdot D^3 \cdot (\vartheta_W - \vartheta_0)}{2^3 \cdot T_0 \cdot v^2} \right)^{1/6} \cdot f_3(Pr) \right]^2 \cdot \frac{2 \cdot \lambda}{\pi \cdot D}$$

Both equations inserted in that of the insulation surface temperature deliver:

$$\vartheta_W = \vartheta_0 + (\vartheta_1 - \vartheta_0) \cdot \frac{1}{1 + \dfrac{D}{2 \cdot \lambda_I} \cdot \ln \dfrac{D \cdot \alpha}{d}}$$

The function $f_3(Pr)$ has the value of 0.83026. The above equation can only be solved with an equation solver or iteration. With *Mathcad* the result is **$D = 461$ mm**.

Discussion

The calculation requires a solver or an iteration procedure.

4.3.2   Sphere

The *Nußelt* and *Rayleigh* number are determined with the diameter of the sphere.

$$Nu_d = 0.56 \cdot \left[Pr / (0.864 + Pr) \cdot Ra \right]^{0.25} + 2 \tag{4.15}$$

4.4   Interaction of free and forced convection

Often free and forced convection are interacting. For this case a combined *Nußelt* number with the *Nußelt* numbers of free and forced convection is reported [4.2]. For parallel-flow and counterflow of the forced convection, different equations are prosed. For the parallel-flow the following equation applies.

$$Nu_l = \sqrt[3]{Nu_{l,forced}^3 + Nu_{l,free}^3} \tag{4.16}$$

For counterflow Equation 4.17 holds:

$$Nu_l = \sqrt[3]{Nu_{l,forced}^3 - Nu_{l,free}^3} \tag{4.17}$$

5   Condensation of pure vapors

When vapor comes into contact with a wall at a temperature below its saturation temperature, condensate is generated on the wall surface. The condensate can either build a film or droplets. A differentiation between *film condensation* and *droplet condensation* has to be made. Droplet condensation reaches higher heat transfer coefficients but requires special dewetting surfaces.

Condensation can occur with pure superheated, saturated or wet vapor but also with gas mixtures like in dew formation. In this book, only the condensation of pure vapors is discussed. At condensation the heat transfer coefficients are a function of the geometry, material properties and the temperature difference between wall and saturation temperature. At high vapor velocities the sheer forces of the vapor flow influence the boundary layer and govern the heat transfer coefficients.

5.1   Film condensation of pure, static vapor

At film condensation of pure, static, saturated vapor a film is generated at the colder wall, which flows downward due to gravity and its thickness increases with flow length. The flow of the film is first laminar and with increasing film thickness turbulent flow occurs. Correlations for laminar and turbulent film condensation as well as for the transition zone will be presented.

During film condensation of static vapor, vapor flows to the wall but its velocity does not influence the heat transfer coefficients.

5.1.1   Laminar film condensation

5.1.1.1   *Condensation of saturated vapor on a vertical wall*

Nußelt [5.1] published in 1916 analytically developed heat transfer coefficients for laminar film condensation on vertical walls with constant wall temperature. He determined the thickness of a condensate film, forced downward by gravity and fed with condensing vapor (water skin theory). The local heat transfer coefficient α_x at location x of the wall is determined by the heat conduction through the film.

$$\alpha_x = \frac{\lambda_l}{\delta_x} \tag{5.1}$$

Therein λ_l is the thermal conductivity of the condensate film and δ_x the film thickness at location x. For the analysis a constant wall temperature ϑ_w is assumed. The vapor has the saturation temperature ϑ_s.

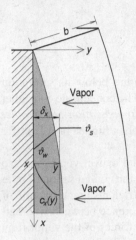

Figure 5.1: Laminar condensation on a vertical wall

Figure 5.1 demonstrates the laminar condensation on a vertical wall. Two forces are acting in the film: The gravity force F_s, generating the downward flow and the frictional force F_τ, acting against the gravity force. As in the film we have a steady state flow, at each location x the temperature and velocity profile are also in steady state.

At the location x in a distance of y from the wall the following gravity force dF_s acts on the mass element dm:

$$dF_s = g \cdot dm = (\rho_l - \rho_g) \cdot b \cdot (\delta_x - y) \cdot g \cdot dx \qquad (5.2)$$

The frictional force dF_τ, generated by the sheer stress τ, at the distance y acting on the mass element is:

$$dF_\tau = \tau \cdot A = \tau \cdot b \cdot dx \qquad (5.3)$$

Both forces have the same magnitude but contrary direction, so we receive:

$$\tau = -(\rho_l - \rho_g) \cdot (\delta_x - y) \cdot g \qquad (5.4)$$

In a laminar flow *Newton* defined the sheer stress τ as:

$$\tau = -\eta_l \cdot \frac{dc_x}{dy} \qquad (5.5)$$

For the velocity gradient at the location x we receive the following differential equation:

$$dc_x = \frac{g \cdot (\rho_l - \rho_g)}{\eta_l} \cdot (\delta_x - y) \cdot dy \tag{5.6}$$

With the boundary condition that at the wall (location $y = 0$) the velocity is zero, the integration delivers:

$$c_x(y) = \frac{g \cdot (\rho_l - \rho_g)}{\eta_l} \cdot (\delta_x \cdot y - \frac{1}{2} y^2) \tag{5.7}$$

The mass flow rate at the location x in the film can be determined by integrating the flow velocity over the cross-section of the film and multiplying the integral with the density of the condensate.

$$\dot{m}_x = \rho_l \cdot b \cdot \int_{y=0}^{y=\delta_x} c_x \cdot dy = \frac{g \cdot (\rho_l - \rho_g) \cdot b}{v_l} \cdot \frac{\delta_x^3}{3} \tag{5.8}$$

Due to the condensing vapor on the film surface $b \cdot dx$ the mass flow rate of the film in the flow direction x increases. From the heat balance and with the rate equation we receive:

$$\delta \dot{Q} = \alpha_x \cdot b \cdot (\vartheta_s - \vartheta_W) \cdot dx = \frac{\lambda_l}{\delta_x} \cdot b \cdot (\vartheta_s - \vartheta_W) \cdot dx = r \cdot d\dot{m}_x \tag{5.9}$$

The change of the mass flow rate is:

$$d\dot{m}_x = \frac{\lambda_l \cdot b \cdot (\vartheta_s - \vartheta_W)}{\delta_x \cdot r} \cdot dx \tag{5.10}$$

The derivation of Equation (5.8) with respect to $d\delta_x$ determine also the change of the mass flow rate.

$$d\dot{m}_x = \frac{b \cdot g \cdot (\rho_l - \rho_g)}{v_l} \delta_x^2 \cdot d\delta_x \tag{5.11}$$

Equations (5.10) and (5.11) deliver following differential equation for the film thickness:

$$\frac{\lambda_l \cdot (\vartheta_s - \vartheta_W) \cdot v_l}{r \cdot g \cdot (\rho_l - \rho_g)} \cdot dx = \delta_x^3 \cdot d\delta_x \tag{5.12}$$

The integration from 0 to x delivers the thickness δ_x of the boundary layer at location x:

$$\delta_x = \left(\frac{4 \cdot \lambda_l \cdot (\vartheta_s - \vartheta_W) \cdot \nu_l}{r \cdot g \cdot (\rho_l - \rho_g)} \cdot x \right)^{0.25}$$

(5.13)

The local heat transfer coefficient α_x at location x:

$$\alpha_x = \frac{\lambda_l}{\delta_x} = \left(\frac{\lambda_l^3 \cdot r \cdot g \cdot (\rho_l - \rho_g)}{4 \cdot (\vartheta_s - \vartheta_W) \cdot \nu_l \cdot l} \right)^{0.25}$$

(5.14)

For the design of heat exchangers not the local but the mean heat transfer coefficient is relevant. The mean heat transfer coefficient of a vertical wall with the length l we receive by integrating equation (5.14) as:

$$\alpha = \frac{1}{l} \cdot \int_{x=0}^{x=l} \alpha_x \cdot dx = 0.943 \cdot \left(\frac{\lambda_l^3 \cdot r \cdot g \cdot (\rho_l - \rho_g)}{(\vartheta_s - \vartheta_W) \cdot \nu_l \cdot x} \right)^{0.25}$$

(5.15)

The latent heat of evaporation r and the vapor density is determined at saturation temperature of the vapor. The material properties of the condensate film have to be determined at the mean temperature $(\vartheta_s + \vartheta_W)/2$ of the film.

Heat transfer coefficients with the index x are local values, that without index are mean values.

EXAMPLE 5.1: Determination of film thickness and heat transfer coefficient

Determine the film thickness and heat transfer coefficient of water and Freon R134a at a vertical wall at $x = 0.1$ m and $x = 1.0$ m. For both fluids the difference between saturation and wall and saturation temperature is 10 K. The material properties are:

	λ W/(m K)	ρ_l kg/m³	ρ_g kg/m³	ν_l 10^6 m²/s	r kJ/kg
Water:	0.682	958.4	0.60	0.295	2 257.9
Freon R134a:	0.094	1 295.2	14.43	0.205	198.6

Solution

Assumptions

- The wall temperature is constant.
- The film flow is laminar.

Analysis

The film thickness is calculated with Equation (5.13), the heat transfer coefficient with Equation (5.14).

$$\delta_x = \left(\frac{4 \cdot \lambda_l \cdot (\vartheta_s - \vartheta_W) \cdot v_l}{r \cdot g \cdot (\rho_l - \rho_g)} \cdot x \right)^{0.25} \qquad \alpha_x = \frac{\lambda_l}{\delta_x}$$

With the given date we receive following values:

	$\delta_{x = 0.1\,m}$ mm	$\alpha_{x = 0.1\,m}$ W/(m^2 K)	$\delta_{x = 1\,m}$ mm	$\alpha_{x = 1\,m}$ W/(m^2 K)
Water:	0.078	8 690	0.140	4 887
Freon R134a:	0.075	1261	0.133	709

Discussion

For both fluids the film thicknesses are rather small and have almost the same value. The lower heat transfer coefficients of Freon 134a results from the lower thermal conductivity of Freon compared to that of water.

Attention! The latent heat of evaporation r must be used with the unit J/kg and not with kJ/kg!

5.1.1.2   *Influence of the changing wall temperature*

In practice, the assumed constant wall temperature would rarely be realized. In most cases the heat released by condensation is absorbed by a fluid and its temperature increases. Is the inlet temperature of the fluid ϑ_1, that at the outlet ϑ_1 and the overall heat transfer coefficient k, the heat rate can be determined.

$$\dot{Q} = k \cdot A \cdot \Delta\vartheta_m = k \cdot A \cdot \frac{\vartheta_1'' - \vartheta_1'}{\ln\left[(\vartheta_s - \vartheta_1') / (\vartheta_s - \vartheta_1'')\right]} \qquad (5.16)$$

The heat rate released by the vapor can be calculated with a mean wall temperature $\overline{\vartheta_W}$.

$$\dot{Q}/A = \dot{q} = \alpha \cdot (\vartheta_s - \overline{\vartheta}_W) \qquad (5.17)$$

With Equations (5.16) and (5.17) we receive for the mean temperature difference between the saturation and wall temperature:

$$(\vartheta_s - \overline{\vartheta}_W) = \Delta\vartheta_m \cdot k / \alpha \qquad (5.18)$$

A large number of tests prove that the mean heat transfer coefficients calculated with the mean temperature difference as given with Equation (5.18) result in high accuracy. The temperature difference can be replaced by the flux $\dot{q}$ in Equation (5.17). For the heat transfer coefficient the result is:

$$\alpha = 0.943 \cdot \left(\frac{\lambda_l^3 \cdot r \cdot g \cdot (\rho_l - \rho_g) \cdot \alpha}{\dot{q} \cdot \nu_l \cdot l} \right)^{1/4} = 0.925 \cdot \left(\frac{\lambda_l^3 \cdot r \cdot g \cdot (\rho_l - \rho_g)}{\dot{q} \cdot \nu_l \cdot l} \right)^{1/3} \quad (5.19)$$

With known dimension of the vertical heat exchanging area A the heat flux $\dot{q} = \dot{Q} / A$ can be replaced by the heat rate. For a rectangular wall the heat exchanging surface area is the product of length l and width b. For a heat exchanger consisting of n vertical tubers the surface area is: $A = b \cdot l = n \cdot \pi \cdot d_a \cdot l$. For tube bundle with vertical tube we receive:

$$\alpha = 0.925 \cdot \left(\frac{\lambda_l^3 \cdot r \cdot g \cdot (\rho_l - \rho_g) \cdot n \cdot \pi \cdot d_a}{\dot{Q} \cdot \nu_l} \right)^{1/3} \quad (5.20)$$

With unknown heat rate, first the heat transfer coefficient is determined with an assumed temperature difference and the final value must be determined by iteration.

5.1.1.3 *Condensation of wet and superheated vapor*

The heat transfer coefficient of condensing static vapor is not influenced by the state of the pure vapor, i.e. the heat transfer coefficients of superheated, saturated and wet vapor depends only on wall and saturation temperature. The only condition is that the wall temperature is lower than the saturation temperature of the vapor.

According to the enthalpy $h_g = h\,(p, \vartheta, x)$ of the vapor, the mass flow rate of condensate and heat rate are determined. According to the energy balance equation the heat rate is:

$$\dot{Q} = \dot{m}_{vapor} \cdot \begin{vmatrix} (h'' - h') = r & \text{saturated vapor} \\ (h(p, \vartheta) - h') & \text{superheated vapor} \\ (x \cdot h'' - h') & \text{wet vapor} \end{vmatrix} \quad (5.21)$$

Wherein h' is the enthalpy of saturated condensate, h'' that of saturated vapor and x the vapor quality. The mass flow rate of the produced condensate is the mass flow rate of vapor multiplied by x, because the wet vapor carries some condensate with it, which will not be condensed.

5.1.1.4   Condensation on inclined walls

The gravity force is reduced on an inclined wall, correspondingly the heat transfer coefficient is reduced with angel φ to the vertical.

$$\alpha = \alpha_{vertical} \cdot (\cos \varphi)^{0.25} \tag{5.22}$$

5.1.1.5   Condensation on horizontal tubes

Most heat exchangers with condensation have horizontal tubes. In the equation for heat transfer coefficients (5.15) the wall length is replaced be the tube diameter.

$$\alpha = 0.728 \cdot \left(\frac{\lambda_l^3 \cdot r \cdot g \cdot (\rho_l - \rho_g)}{(\vartheta_s - \bar{\vartheta}_w) \cdot \nu_l \cdot d_a} \right)^{0.25} \tag{5.23}$$

Similarly as for a vertical wall the temperature difference can be replaced by the heat flux and heat transfer coefficient. The heat transfer surface area is $A = n \cdot \pi \cdot d_a$.

$$\alpha = 0.959 \cdot \left(\frac{\lambda_l^3 \cdot r \cdot g \cdot (\rho_l - \rho_g) \cdot n \cdot l}{\dot{Q} \cdot \nu_l} \right)^{1/3} \tag{5.24}$$

5.1.2   Turbulent film condensation on vertical surfaces

On longer walls the thickness of the condensate film increases with the flow length and the flow transits from laminar to turbulent. Analytic determination of the heat transfer coefficients is not possible for the turbulent flow. The transition zone from laminar to turbulent and fully turbulent condensation will be discussed with the introduction of dimensionless similarity numbers.

5.2   Dimensionless similarity numbers

Similar to the convective heat transfer, those of condensation will be represented with the *Nußelt* number as a function of dimensionless numbers [5.2]. For the *Nußelt* number the characteristic length L is defined as:

$$L' = \sqrt[3]{\frac{\nu_l^2}{g}} \tag{5.25}$$

Nußelt number:

$$Nu_{L'} = \frac{\alpha \cdot L'}{\lambda_l} \tag{5.26}$$

The second similarity number is the *Reynolds* number. It is defined as:

$$Re_l = \frac{\Gamma}{\eta_l} \qquad (5.27)$$

The parameter Γ is the *mass flow rate per unit width*. It is the mass flow rate of produced condensate per width b of a vertical surface area, which is called the *discharge width*.

$$\Gamma = \frac{\dot{m}_l}{b} \qquad (5.28)$$

The mass flow rate of produced condensate multiplied by the latent heat of evaporation is the heat rate. The mass flow rate per unit width can also be given as a function of the heat rate.

$$\Gamma = \frac{\dot{m}_l}{b} = \frac{\dot{Q}}{r \cdot b} \qquad (5.29)$$

The discharge width b is defined:

at vertical walls as the width of the wall: $b = b$
at vertical tubes as the sum of all tube circumferences: $b = n \cdot \pi \cdot d$

With this similarity numbers and definitions the local and mean *Nußelt* number can be calculated.

5.2.1   Local heat transfer coefficients

The local *Nußelt* number for laminar flow is determined by inserting the dimensionless similarity numbers into Equations (5.14), (5.17) and (5.29):

$$Nu_{L', \, lam, \, x} = \frac{\alpha_x \cdot L'}{\lambda_l} = 0.693 \cdot \left(\frac{1 - \rho_g / \rho_l}{Re_l} \right)^{1/3} \cdot f_{wave} \qquad (5.30)$$

Herein f_{wave} is a correction function taking into account the influence of the waviness of the film surface at higher *Reynolds* numbers. It is defined as:

$$f_{wave} = \begin{cases} 1 & \text{für} & Re_l < 1 \\ Re_l^{0.04} & \text{für} & Re_l \geq 1 \end{cases} \qquad (5.31)$$

For the local turbulent *Nußelt* number the following empirical correlation holds:

$$Nu_{L', \, turb, \, x} = \frac{\alpha_x \cdot L'}{\lambda_l} = \frac{0.0283 \cdot Re_l^{7/24} \cdot Pr_l^{1/3}}{1 + 9.66 \cdot Re_l^{-3/8} Pr_l^{-1/6}} \qquad (5.32)$$

To cover the whole range of laminar and turbulent condensation as well as the transit zone, the following equation was found, which is in best conformity with test results:

$$Nu_{L',x} = \frac{\alpha_x \cdot L'}{\lambda_l} = \sqrt{Nu_{L',lam}^2 + Nu_{L',turb}^2} \cdot f_\eta \qquad (5.33)$$

The correction function f_η treats the temperature dependence of the viscosity.

$$f_\eta = (\eta_{ls} / \eta_{lW})^{0.25} \qquad (5.34)$$

5.2.2   Mean heat transfer coefficients

The mean heat transfer coefficient and *Nußelt* number are determined by integration over the length of the heat transfer surface area.

$$Nu_{L',lam} = \frac{\alpha \cdot L'}{\lambda_l} = 0.925 \cdot \left(\frac{(1 - \rho_g / \rho_l)}{Re_l} \right)^{1/3} \cdot f_{wave} \qquad (5.35)$$

$$Nu_{L',turb} = \frac{\alpha \cdot L'}{\lambda_l} = \frac{0.020 \cdot Re_l^{7/24} \cdot Pr_l^{1/3}}{1 + 20.52 \cdot Re_l^{-3/8} Pr_l^{-1/6}} \qquad (5.36)$$

$$Nu_{L'} = \frac{\alpha \cdot L'}{\lambda_l} = {}^{1.2}\!\sqrt{Nu_{L',lam}^{1.2} + Nu_{L',turb}^{1.2}} \cdot f_\eta \qquad (5.37)$$

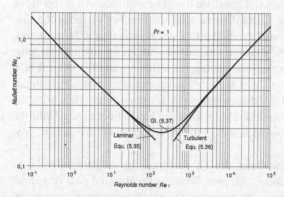

Figure 5.2: Mean *Nußelt* number of condensing pure static vapor

5.2.3   Condensation on horizontal tubes

For condensation on horizontal tubes the similarity numbers inserted in Equation (5.24) result in the mean *Nußelt* number:

$$Nu_{L'} = 0.959 \cdot \left(\frac{1 - \rho_g / \rho_l}{Re_l} \right)^{1/3}$$

(5.38)

We have to pay attention to the fact that the discharge width b has to be determined with tube length l, i.e. $b = n \cdot l$.

5.2.4   Procedure for the determination of heat transfer coefficients

In most cases heat exchangers with condensation are of the tube and shell type. This chapter is limited to the discussion of apparatus in which condensation happens on the outer surface of vertical and horizontal tubes. It has to be distinguished between design and recalculation procedures.

For a design the thermal and also some geometry data (heat rate, saturation temperature, vapor properties, tube diameter and materials) are given. The heat transfer coefficient of the fluid to which the heat is transferred has to be known or calculated. For these data the heat exchanger must be designed, i.e. the number of tubes and their length is to be determined, such that the given thermal data can be reached with a certain plausibility. Unknowns are: the number of tubes, length of tubes, difference between saturation and wall temperature and the mean temperature for the determination of the material properties. A part of these unknowns has first to be assumed and with iteration the calculation procedures must be repeated until the required accuracy is reached. Furthermore, for example, the tube diameter may be subject to optimization. In such a case the calculation for several diameters has to be performed and the different solutions evaluated economically. The number of tubes influences the condensation heat transfer coefficients but is normally determined by the required flow velocity in the tubes. Therefore it will be assumed as a given value as well as the tube diameter. For vertical rows with the number of tubes the discharge width is known, for horizontal tubes the tube length has first to be assumed. The material properties in the first step are calculated with an assumed mean temperature. With these assumptions the heat transfer coefficients and the overall heat transfer coefficient, the tube length and the wall temperature can be determined. Now the material properties and the discharge width for horizontal tubes with the received tube length can be determined. This procedure has to be repeated until the required accuracy is reached. Figure 5.3 shows the flow diagram of the design procedure.

The recalculation is used for condenser already designed, i.e. the geometrical data of the tubes as well as their number is known. Some of the thermal data (e.g. heat transfer coefficients, mass flow rates, vapor properties and inlet temperature) of the fluid in the tubes and also that of the vapor may change. For the changed conditions the change of the other thermal data has to determined.

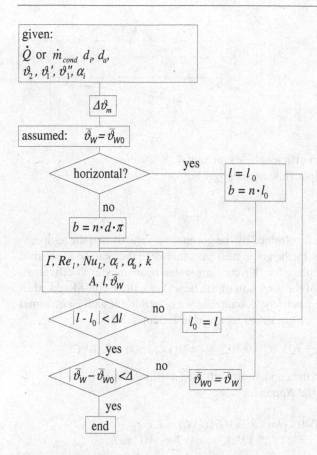

Figure 5.3: Flow diagram for the design calculation procedure

At recalculations, for example, for a changed heat rate, the outlet temperature of the cooling fluid can be determined. For the determination of the material properties first a saturation temperature must be assumed. With it inside and outside heat transfer coefficients, the overall heat transfer coefficient can be calculated and also the required log mean temperature difference. From the latter the saturation temperature and wall temperature can be determined.

EXAMPLE 5.2: Design of a power plant condenser

The condensation heat transfer coefficient of the condenser from Example 3.3 was assumed. Now it can be calculated.

Find

The required length of the condenser tubes.

Solution

Schematic   See sketch.

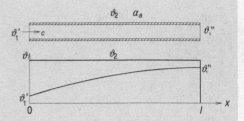

Assumptions

- The mean heat transfer coefficients are constant.
- The influences of the vapor flow are negligible.

Analysis

The number of tubes were determined by the given flow velocity and the temperature rise of the cooling water by the given heat rate, therefore they can be used from Example 3.3. The same applies for the cooling water heat transfer coefficients, whereas here the influence of the direction of the heat flux will be considered additionally. The material properties of the condensate we determine at with an assumed mean temperature, with the temperatures received in Example 3.3:

$$\bar{\vartheta}_W = \vartheta_2 - \Delta\vartheta_m \cdot k / \alpha_a = 35 \ °C - 9.102 \cdot K \cdot 4403 / 13500 = 32.0 \ °C$$

The material properties of the condensate will be determined at 33.5 °C. The values from the vapor table in the Appendix A6:

$$\rho_l = 995.5 \ kg/m^3, \ \rho_g = 0.040 \ kg/m^3, \ \lambda_l = 0.616 \ W/(m \ K),$$
$$\eta_l = 792.2 \cdot 10^{-6} \ kg/(m \ s), \ r = 2417.9 \ kJ/kg, \ \nu_l = 0.796 \cdot 10^{-6} \ m^2/s.$$

The mass flow rate per unit width we calculate with Equation (5.29):

$$\Gamma = \frac{\dot{Q}}{r \cdot n \cdot l} = \frac{2\,000 \cdot 10^6 \cdot W \cdot kg}{2\,417\,900 \ \cdot J \cdot 57\,727 \cdot 11.562 \cdot m} = 1.239 \cdot 10^{-3} \ \frac{kg}{m \cdot s}$$

Reynolds number (5.27): $Re_l = \dfrac{\Gamma}{\eta_l} = \dfrac{1.239 \cdot 10^{-3}}{792.2 \cdot 10^{-6}} / = 1.564$

For the *Nußelt* number Equation (5.38) delivers:

$$Nu_{L'} = 0.959 \cdot \left(\frac{1 - \rho_g / \rho_l}{Re_l}\right)^{1/3} = 0.959 \cdot \left(\frac{1 - 0.04 / 995.5}{1.578}\right)^{1/3} = 0.8261$$

For the determination of the heat transfer coefficient we need the characteristic length L given with Equation (5.25).

$$L' = \sqrt[3]{\nu_l^2 / g} = \sqrt[3]{(0.796 \cdot 10^{-6})^2 \cdot m^4 \cdot s^2 / (9.806 \cdot m \cdot s^2)} = 0.04013 \cdot 10^{-3} \ m$$

The heat transfer coefficient is:

$$\alpha = Nu_{L'} \cdot \lambda / L' = 0.8238 \cdot 0.616 \cdot W/(0.04013 \cdot 10^{-3} \cdot m \cdot K \cdot m) = 12\,682 \ W/(m^2 \cdot K)$$

It is somewhat smaller than the value of 13500 W/(m² K), used in Example 3.3.

The heat transfer coefficient of the cooling water in Example 3.3 was calculated as 8481 W/(m² K). For the overall heat transfer coefficient we receive:

$$k = \left(\frac{1}{\alpha_a} + \frac{d_a}{2 \cdot \lambda_R} \cdot \ln \frac{d_a}{d_i} + \frac{d_a}{d_i \cdot \alpha_i} \right)^{-1} =$$

$$= \left(\frac{1}{12\,682} + \frac{0.024}{2 \cdot 17} \cdot \ln \frac{24}{23} + \frac{24}{23 \cdot 8\,481} \right)^{-1} = 4\,273 \ \frac{W}{m^2 \cdot K}$$

The overall heat transfer coefficient is smaller than calculated in Example 3.3, consequently the tube length will be proportionally larger with 11.803 m. The tube length increases in proportion to the decrease of the overall heat transfer coefficient. For the iteration it is advisable to use the correction function f_2 for the tube side heat transfer coefficient. With the correction function the equation for the overall heat transfer coefficient is:

$$k = \left(\frac{1}{\alpha_a} + \frac{d_a}{2 \cdot \lambda_R} \cdot \ln \frac{d_a}{d_i} + \frac{d_a}{d_i \cdot \alpha_i \cdot f_1 \cdot f_2} \right)$$

The *Prandtl* number of the cooling water at 30 °C has the value of 5.414. The wall temperature inside the tube is:

$$\vartheta_{Wi} = \vartheta_m + \Delta \vartheta_m \cdot k / \alpha_i$$

After the fifth repetition of the calculations the following values were found:

$$f_1 = 1.016, f_2 = 1.015, \ \alpha_a = 12700 \ W/(m^2 \ K), \ k = 4\,348 \ W/(m^2 \ K),$$
$$\mathbf{l = 11.611 \ m}$$

The change of the material properties of the condensate was taken into account.

Discussion

The correction functions f_1 and f_2 are responsible for only a small change of 0.5 %. The design of a large condenser requires a high accuracy, because on the one hand the size of the heat transfer area determines the price and on the other hand in case of not reaching the guaranteed values, high commercial penalties are the consequence.

EXAMPLE 5.3: Design and recalculation of a condenser for Freon R134a

On the vertical tubes of a condenser 0.5 kg/s Freon R134a should be condensed at 50 °C. In the tubes water flows with a velocity of 1 m/s. The cooling water shall be heated from 40 to 45 °C. The copper tubes have an outer diameter of 12 mm and wall thickness of 1 mm; the thermal conductivity is 372 W/(m² K).
  Material properties of the cooling water at 42.5 °C:
  $\rho = 991.3$ kg/m³, $c_p = 4.178$ kJ/(kg K), $\lambda = 0.634$ W/(m K), $v = 0.629 \cdot 10^{-6}$ m²/s, $Pr = 4.1$.
  Material properties of Freon 134a:
  $\rho_l = 1\,102.3$ kg/m³, $\rho_g = 66.3$ kg/m³, $\lambda = 0.071$ W/(m K), hl $= 142.7 \cdot 10^{-6}$ kg/(m s), $r = 151.8$ kJ/kg, $v_l = 0.132 \cdot 10^{-6}$ m²/s, $Pr_l = 3.14$.
  For simplification of the calculations it can be assumed that the influence of the tube length and the direction of the heat flux are negligible.

Find

a) Number and length of the tubes.
b) Heat rate and heat transfer coefficients for an increased vapor mass flow rate of 0.65 kg/s. The change of material properties is negligeable.

Solution

Schematic   Temperature gradient as in Example 5.2

Assumptions

• The mean heat transfer coefficients are constant.
• The influence of tube length and heat flux direction are negligible.
• The change of material properties in part b) is negligible.
• The vapor velocity has no influence on heat transfer.

Analysis

a)   The mass flow rate of cooling water is calculated from the energy balance equation of Freon and cooling water.

$$\dot{Q} = \dot{m}_{R134a} \cdot r \qquad\qquad \dot{Q} = \dot{m}_{CW} \cdot c_p \cdot (\vartheta_1'' - \vartheta_1')$$

$$m_{CW} = \frac{\dot{m}_{R134a} \cdot r}{c_p \cdot (\vartheta_1'' - \vartheta_1')} = \frac{0.5 \cdot \text{kg} \cdot 151\,800 \cdot \text{J} \cdot \text{kg} \cdot K}{4\,178 \cdot \text{J} \cdot 5 \cdot \text{K} \cdot \text{kg} \cdot \text{s}} = 3.633\,\frac{\text{kg}}{\text{s}}$$

The number of tubes is determined with the given water velocity.

$$n = \frac{4 \cdot \dot{m}_{CW}}{c_{CW} \cdot \rho_{CW} \cdot \pi \cdot d_i^2} = \frac{4 \cdot 3.633 \cdot \text{kg} \cdot \text{s} \cdot \text{m}^3}{1 \cdot \text{m} \cdot 991.3 \cdot \text{kg} \cdot \pi \cdot 0.01^2 \cdot \text{m}^2 \cdot \text{s}} = 47$$

The wetting density is calculated with Equation (5.29).

$$\Gamma = \frac{m_{R134a}}{n \cdot \pi \cdot d_a} = \frac{0.5 \cdot \text{kg}}{47 \cdot \pi \cdot 0.012 \cdot \text{m} \cdot \text{s}} = 0.2822 \frac{\text{kg}}{\text{m} \cdot \text{s}}$$

Reynolds number (5.27): $Re_l = \Gamma / \eta_l = 0.2822 / 142.7 \cdot 10^{-6} = 1978$

The *Nußelt* number is calculated with Equations (5.31) to (5.37).

$$f_{wave} = Re_l^{0.04} = 1.355$$

$$Nu_{L', lam} = 0.925 \cdot \left(\frac{(1 - \rho_g / \rho_l)}{Re_l} \right)^{1/3} \cdot f_{wave} = 0.925 \cdot \left(\frac{(1 - 66.3/1102.3)}{1978} \right)^{1/3} \cdot 1.355 = 0.0978$$

$$Nu_{L', turb} = \frac{0.020 \cdot Re_l^{7/24} \cdot Pr_l^{1/3}}{1 + 20.52 \cdot Re_l^{-3/8} Pr_l^{-1/6}} = \frac{0.020 \cdot 1978^{7/24} \cdot 3.14^{1/3}}{1 + 20.52 \cdot 1978^{-3/8} 3.14^{-1/6}} = 0.135$$

$$Nu_{L'} = \frac{\alpha \cdot L'}{\lambda_1} = \sqrt[1.2]{(Nu_{L', lam})^{1.2} + Nu_{L', turb}^{1.2}} = 0.2079$$

The characteristic length L is calculated with Equation (5.25).

$$L' = \sqrt[3]{v_l^2 / g} = \sqrt[3]{(0.132 \cdot 10^{-6})^2 \cdot \text{m}^4 \cdot \text{s}^2 / (9.806 \cdot \text{m} \cdot \text{s}^2)} = 0.01211 \cdot 10^{-3} \text{ m}$$

The heat transfer coefficient is:

$$\alpha = Nu_{L'} \cdot \lambda / L' = 0.2079 \cdot 0.071 \cdot \text{W} / (0.01211 \cdot 10^{-3} \cdot \text{m} \cdot \text{K} \cdot \text{m}) = 1219 \text{ W} / (\text{m}^2 \cdot \text{K})$$

The heat transfer coefficient is calculated with Equation (3.8).

$$Re_{d_i} = \frac{c_1 \cdot d_i}{v_1} = \frac{1 \cdot 0.01}{0.629 \cdot 10^{-6}} = 15898$$

$$\xi = [1.8 \cdot \log(Re_{d_i}) - 1.5]^{-2} = 0.0272$$

$$Nu_{d_i} = \frac{\xi / 8 \cdot Re \cdot Pr}{1 + 12.7 \cdot \sqrt{\xi / 8} \cdot (Pr^{2/3} - 1)} = 102.8$$

$$\alpha_i = Nu_{d_i} \cdot \lambda_1 / d_i = 6517 \text{ W} / (\text{m}^2 \cdot \text{K})$$

The overall heat transfer coefficient is:

$$k = \left(\frac{1}{\alpha} + \frac{d_a}{2 \cdot \lambda_R} \cdot \ln \frac{d_a}{d_i} + \frac{d_a}{d_i \cdot \alpha_i} \right)^{-1} =$$

$$= \left(\frac{1}{1219} + \frac{0.012}{2 \cdot 372} \cdot \ln \frac{12}{10} + \frac{12}{10 \cdot 6517} \right)^{-1} = 992.6 \ \frac{W}{m^2 \cdot K}$$

The log mean temperature difference:

$$\Delta \vartheta_m = \frac{\vartheta_1'' - \vartheta_1'}{\ln \left(\dfrac{\vartheta_2 - \vartheta_1'}{\vartheta_2 - \vartheta_1''} \right)} = \frac{(45 - 40) \cdot K}{\ln \left(\dfrac{50 - 40}{50 - 45} \right)} = 7.213 \ K$$

For the required tube length we receive:

$$l = \frac{\dot{Q}}{k \cdot \Delta \vartheta_m \cdot n \cdot \pi \cdot d_a} = \frac{75\,900 \ \cdot W \cdot m^2 \cdot K}{992.6 \cdot W \cdot 7.123 \ \cdot K \cdot 47 \cdot \pi \cdot 0.012 \cdot m} = \mathbf{5.983 \ m}$$

b)   Due to the assumption of constant material properties the heat transfer coefficient in the tubes does not change. The outlet temperature of the cooling water rises by 6.5 K, i.e. the outlet temperature is now 46.5 °C. With the increased vapor mass flow rate we receive for condensation and overall heat transfer:

$$\Gamma = \frac{m_{R134a}}{n \cdot \pi \cdot d_a} = \frac{0.65 \cdot kg}{47 \cdot \pi \cdot 0.012 \cdot m \cdot s} = 0.3668 \ \frac{kg}{m \cdot s}$$

$$Re_l = \Gamma / \eta_l = 0.3668 / 142.7 \cdot 10^{-6} = 2571$$

$$f_{wave} = Re_l^{0.04} = 1.369$$

$$Nu_{L', \, lam} = 0.925 \cdot \left(\frac{(1 - \rho_g / \rho_l)}{Re_l} \right)^{1/3} \cdot f_{wave} = 0.0905$$

$$Nu_{L', \, turb} = \frac{0.020 \cdot Re_l^{7/24} \cdot Pr_l^{1/3}}{1 + 20.52 \cdot Re_l^{-3/8} Pr_l^{-1/6}} = 0.1528$$

$$Nu_{L'} = \frac{\alpha \cdot L'}{\lambda_l} = \sqrt[1.2]{(Nu_{L', \, lam})^{1.2} + Nu_{L', \, turb}^{1.2}} = 0.2138$$

$$\alpha = \frac{Nu_{L'} \cdot \lambda_2}{L'} = 1279.5 \ \frac{W}{m^2 \cdot K}$$

$$k = \left(\frac{1}{\alpha} + \frac{d_a}{2 \cdot \lambda_R} \cdot \ln \frac{d_a}{d_i} + \frac{d_a}{d_i \cdot \alpha_i} \right)^{-1} =$$

$$= \left(\frac{1}{1279.5} + \frac{0.012}{2 \cdot 372} \cdot \ln \frac{12}{10} + \frac{12}{10 \cdot 6517} \right)^{-1} = 1032 \ \frac{W}{m^2 \cdot K}$$

To be able to discharge the higher heat rate, the log mean temperature difference increases. Its value can be calculated with Equation (5.16).

$$\Delta \vartheta_m = \frac{\dot{Q}}{k \cdot A} = \frac{\dot{Q}}{k \cdot n \cdot \pi \cdot d_a \cdot l} = \frac{98670 \ \cdot W \cdot m^2 \cdot K}{1032 \ \cdot W \cdot 47 \cdot \pi \cdot 0{,}012 \cdot m \cdot 5.983 \cdot m} = 9{,}048 \ K$$

The log mean temperature difference delivers the condensation temperature:

$$\Theta = e^{\frac{\vartheta_1'' - \vartheta'}{\Delta \vartheta_{ml}}} = 2.0564 \qquad \frac{\vartheta_2 - \vartheta_1'}{\vartheta_2 - \vartheta_1''} = \Theta \qquad \vartheta_2 = \frac{\vartheta_1' - \Theta \cdot \vartheta_1''}{1 - \Theta} = 52.65 \ °C$$

Discussion

The design was performed with given material properties and needed therefore no iteration. Taking into account the influence of the tube length and the material properties for both procedures iterations would be required.

5.2.5   Pressure drop in tube bundles

In tube bundles with horizontal tubes condensate drops from upper tubes and influences the film thickness of the tube underneath. Furthermore the sheer stress forces, caused by vapor flow through the bundle has influence on the condensate film thickness. *Nußelt* developed theoretical correlations for a horizontal tube row with tubes arranged vertically one upon the other. According to this theory, the heat transfer coefficients decrease proportionally to the fourth root of the number of tubes ($n^{-1/4}$). The reduction is due to the increasing condensate film thickness caused by the condensate dropping from the upper tubes. In many publications this law is proposed for tube bundles with several tube rows. Tests, especially with a larger number of tubes, showed that the reduction of the heat transfer coefficients is compensated by the vapor velocity, thus the law for one tube can also be applied for tube bundles.

In large condenser bundles the vapor side pressure drop and accumulation of noncondensable gases have an important influence. This was experienced with large turbine condensers. In steam power plants with an electrical output of approximately 100 MW, condensers with circular arranged tubes in circular shells were used successfully. In the late 1960 and in 1970 the electric power of steam power stations raised over 1000 MW. The experience was made that the circular condensers have not reached the design value, i.e. the pressure was considerably higher than the design value. With increasing bundle diameter the number of tubes and with them the surface area increases by the square but the circumferences only linearly. The steam velocities entering the bundles on the circumference of the bundle increased and with them the pressure drop. The pressure and consequently also the saturation temperature in the bundle were lower than that outside the bundle. To be able to condense all the steam discharged from the turbine the saturation temperature and with it the saturation pressure increase and deteriorate the turbine effectiveness. The improvement measure, installing steam lanes in the bundle to enlarge the circumference failed as in the branches between the lanes non-condensables were collected and these areas were lost for the heat transfer.

On the flow path into a bundle by passing the tube rows, the mass flow rate of the steam decreases due to the condensation and also the velocities decrease. At the circumference of the bundle the largest velocities occur and thus also the largest pressure drops. The tubes in the bundle see a lower pressure, subsequently the saturation temperature lowers, resulting in a lower log mean temperature difference in the bundle. To be able to condense all the steam, the saturation temperature increases. To establish a certain saturation pressure and temperature outside the bundle, which is relevant for the turbine, a larger surface has to be installed. In literature no useful publication exists which describes how to calculate the correct tube bundle arrangement. The *condenser bundle design* is the know-how of the manufacturers.

Example 5.4 demonstrates the effect of pressure drop in simple model.

Tube arrangements of large condensers are established by model test and 3D-flow modelling. Basically the circumference of the bundle is designed so that the steam velocity in not to high to keep the pressure drop at acceptable low levels. Further the bundle design should be such that only one isobar exists, in which the noncompensable can be removed.

EXAMPLE 5.4: Influence of pressure drop on heat flux and surface area

The aim of this example is to show with a very simple model the influence of pressure drop on heat flux as a function of the number of tubes of a circular condenser bundle. The tube bundle has n tubes arranged in a $60°$ net of circular shape. The tubes have 24 mm outer diameter and a distance of 32 mm. They should be arranged as close as possible to the circular shape. The heat transfer coefficient is assumed as constant with 3500 W/(m² K). The inlet temperature of the cooling water is 20 °C and the outlet 30 °C. The saturation temperature outside the bundle is 35 °C. In the model it is assumed that there is pressure drop of the vapor flow only in the circumference tube row and for all the inner tubes of the bundle the saturation pressure and temperature is constant. The hydraulic resistance coefficient between the tubes related to the velocity in the periphery tube gaps is 1.5.

Find

The required additional surface to keep the saturation temperature outside the bundle at 35 °C.

Solution

Schematic   See sketch.

Bundle radius

32 mm

ø24 mm

Assumptions

- The heat transfer coefficient is constant.
- Pressure drop occurs only in the first row.
- The tubes are inserted as close as possible into the circular area.

Analysis

To calculate the area of the bundle first the area A_1, required for one tube, has to be determined. One tube requires the area of two 60° triangles with the side length $of \, s_1$.

$$A_1 = \frac{\sqrt{3}}{2} \cdot s_1^2 = \frac{\sqrt{3}}{2} \cdot 32^2 \cdot mm^2 = 0.887 \cdot 10^{-3} \; m^2$$

The bundle radius required for n tubes is:

$$R = \sqrt{\frac{n}{\pi} \cdot A_1} = \sqrt{\frac{n}{\pi} \cdot \frac{\sqrt{3}}{2}} \cdot s_1 = \sqrt{n} \cdot \sqrt{\frac{\sqrt{3}}{2 \cdot \pi}} \cdot s_1 = \sqrt{n} \cdot 16.801 \; mm$$

The steam flows into the bundle between the periphery tubes. The flow cross-section A_0 between the tubes is:

$$A_0 = \frac{U \cdot l}{s_1} \cdot (s_1 - d) = 2 \cdot \pi \cdot R \cdot l \cdot (1 - d / s_1) = \sqrt{n} \cdot l \cdot 0.026391 \; m$$

To determine the steam volume flow rate, flow velocity, pressure drop and saturation temperature, the material properties of vapor are required.

$\vartheta_s = 35 \; °C$, $p_s = 56.36 \; mbar$, $r = 2418 \; kJ/kg$, $\rho = 0.03961 \; kg/m^3$

The steam volume flow rate relevant for the design is defined by the heat transfer coefficient and log mean temperature at 35 °C saturation temperature.

$$\Delta\vartheta_m = \frac{\vartheta_2 - \vartheta_1}{\ln\left(\dfrac{\vartheta_s - \vartheta_1}{\vartheta_s - \vartheta_2}\right)} = \frac{30 - 20}{\ln(15/5)}\,K = 9.102\;K$$

$$\dot{V}_D = \frac{\dot{m}_D}{\rho} = \frac{\dot{Q}}{r \cdot \rho} = \frac{k \cdot A \cdot \Delta\vartheta_m}{r \cdot \rho} = \frac{k \cdot n \cdot l \cdot \pi \cdot d \cdot \Delta\vartheta_m}{r \cdot \rho} =$$

$$= \frac{n \cdot l \cdot \pi \cdot 3.5 \cdot 0.024 \cdot 9.102 \cdot m^2}{2\,418 \cdot 0.03961 \cdot s} = n \cdot l \cdot 0.02508 \; m^2/s$$

One part of the volume flow rate condenses at outer the tube half at 35 °C saturation temperature. The remaining part condenses in the bundle at a lower saturation temperature, caused be the pressure drop. This part is:

$$\dot{V}_{Din} = \dot{V}_D \cdot [1 - U/(2 \cdot s_1 \cdot n)] = \dot{V}_D \cdot [1 - 2 \cdot \pi \cdot \sqrt{n} \cdot 16.801\;mm/(2 \cdot s_1 \cdot n)] =$$

$$= n \cdot l \cdot 0.02508 \cdot [1 - 1.6494/\sqrt{n}]\;m^2/s$$

The flow velocity between the tubes is:

$$c = \frac{\dot{V}_{Din}}{A_0} = \frac{n \cdot l \cdot 0.02508 \cdot [1 - 1.237/\sqrt{n}]\;m^2/s}{\sqrt{n} \cdot l \cdot 0.019794\;m} = 0.950 \cdot [\sqrt{n} - 1.237]\;m/s$$

The pressure drop:

$$\Delta p = \zeta \cdot c^2 \cdot \rho/2 = 0.75 \cdot 0.9025 \cdot [\sqrt{n} - 1.6494]^2 \cdot 0.03961\;Pa =$$

$$= 0.04769 \cdot [\sqrt{n} - 1.649]^2 Pa \approx 0.02681 \cdot n\;Pa$$

With the pressure drop the saturation temperature can determined in the steam tables. With the pressure decrease, caused by the pressure drop, the log mean temperature decreases so does the heat flux.

$$\dot{q}(n) = k \cdot \Delta\vartheta_m(n)$$

As the heat rate remains constant, the surface area inside the bundle must be increased proportionally to the decrease in heat flux. The changes in percents are:

$$\left(\frac{A}{A_0} - 1\right) \cdot 100\;\% = \left(\frac{\dot{q}_0}{\dot{q}} - 1\right) \cdot 100\;\% = \left(\frac{\Delta\vartheta_{m0}}{\Delta\vartheta_m} - 1\right) \cdot 100\;\%$$

The calculated values are in the table below.

n	Δp	p_s	ϑ_s	$\Delta\vartheta_m$	$\Delta A/A \cdot 100$
-	Pa	mbar	°C	K	%
0	0.0	56.29	35.00	9.10	0.0
2000	49.8	55.79	34.84	8.93	2.0
4000	101.8	55.27	34.67	8.74	4.2
6000	154.2	54.75	34.50	8.55	6.5
8000	206.8	54.22	34.33	8.35	9.0
10000	259.5	53.70	34.15	8.15	11.6
12000	312.3	53.17	33.97	7.95	14.5
14000	365.2	52.64	33.79	7.75	17.5
16000	418.1	52.11	33.61	7.54	20.7
18000	471.1	51.58	33.43	7.33	24.2
20000	524.1	51.05	33.25	7.11	28.0

In this very simple model we do not consider the fact, that with the reduced saturation temperature the warming of the cooling water decreases. This is contrary to the decrease of the log mean temperature difference, which will be some less than calculated. This error is more than compensated by the assumption that the pressure drop only happens in the gap between the periphery tubes. In reality even larger pressure drops were measured.

Discussion

The result of calculations shows that in large tube bundles the pressure drop causes a requirement of additional surface areas. The experience is, that for smaller bundles (less than 2000 tubes) the pressure drop can be neglected. Manufactures of large condensers develop their bundle design by flow test and 3-D computer codes.

5.3   Condensation of pure vapor in tube flow

According to the flow direction of vapor and condensate (parallel- or counterflow), vapor velocity and orientation of the pipe (horizontal or vertical) different equations for the heat transfer coefficients are required. At condensation of pure vapor in a pipe with increasing length, the mass flow rate and with it the inlet velocity of the vapor is increasing, because of the fact, that with increasing pipe length the surface area and the condensate production also increase. With the vapor flow an additional force, caused by the sheer stress of the vapor flow, is acting on the condensate surface. At high vapor flow velocities, the influence of the gravity becomes negligible compared to the sheer stress forces. In vertical tubes the influence of the vapor flow is different for up- and downward vapor flow. Therefore different correlations are required for the flow patterns and pipe orientation.

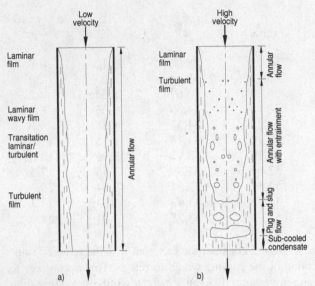

Figure 5.4: Condensation in a tube at downward vapor flow [5.7]:
a) separated flow b) flow with entrainment and flooding

At small vapor velocities in vertical tubes a condensate film can be built over the whole tube length (separated flow); at higher flow velocities droplets (entrainment) can be thrown off the film and after a certain flow length the tube can be filled completely with condensate (Figure 5.4). At upward vapor flow with smaller vapor velocity a condensate film which flows contrary to the stream flow can occur. At higher vapor velocity the vapor flow can transfer the condensate film upward and entrainment and flooding my occur. In this case at the tube end sufficient excess volume must remain, otherwise the condensate reverses the flow and slug and plug flow occurs. With these flow types no steady operation is possible and due to condensate sub-cooling the tube can be subject to high cycle fatigue with consequent damage of the hat exchanger.

In very short tubes, which have low vapor mass flow rates and velocities the correlations as given with Equation (5.38) for horizontal tubes and Equation (5.37) for vertical tubes can be applied, as long as they result in larger values than the correlations proposed in the following chapters.

5.3.1   Condensation in vertical tubes

By the vapor flow a sheer stress acts additionally on the surface of the condensate film [5.3, 5.4]. For determination of the influence of the vapor flow between parallel- and counterflow must be distinguished.

5.3.1.1 Parallel-flow (vapor flow downward)

Vapor enters the upper inlet of a vertical tube and increases the velocity of the downward moving condensate film on the wall. The sheer force applied on the film surface influences the velocity profile in the film as well as the generation of turbulence in the film. Depending on the vapor velocity also droplets and waves can be generated. For the local *Nußelt* number the following correlation is proposed:

$$Nu^*_{L', x} = (1+\tau^*_{ZP})^{1/3} \cdot \sqrt{(C_{lam} \cdot Nu_{L', lam, x})^2 + (C_{turb} \cdot Nu_{L', turb, x})^2} \qquad (5.39)$$

The correction terms C_{lam} and C_{turb} take into account the influence of the vapor flow on laminar and turbulent heat transfer. The *Nußelt* numbers are calculated with Equations (5.30) and (5.32). The dimensionless sheer stress of the two-phase flow τ_{ZP}^* is a function of the dimensionless sheer stress of the pure gas flow in an empty tube τ_g^*.

$$\tau_g^* = \frac{\tau_g}{g \cdot \rho_l \cdot \delta^+} \qquad \text{with} \qquad \tau_g = \frac{\zeta_g \cdot \rho_g \cdot \overline{c}_g^2}{8} \qquad (5.40)$$

The sheer stress τ_g is that of the gas phase in an empty tube, ζ_g the hydraulic resistance coefficient of the vapor, δ^+ the film thickness and $\overline{c}_g$ the mean velocity of the vapor in an empty tube.

$$\overline{c}_g = \frac{4 \cdot \dot{m}_g}{\rho_g \cdot n \cdot \pi \cdot d_i^2} = \frac{4 \cdot \dot{m} \cdot x}{\rho_g \cdot n \cdot \pi \cdot d_i^2} \qquad (5.41)$$

The hydraulic resistance coefficient is determined with the *Reynolds* number of the gas phase.

$$\zeta_g = 0.184 \cdot Re_g^{-0.2} \qquad \text{with} \qquad Re_g = \frac{\overline{c}_g \cdot d}{\nu_g} \qquad (5.42)$$

The dimensionless sheer stress of the two phase flow τ_{ZP}^* is defined as:

$$\tau_{ZP}^* = \tau_g^* \cdot [1 + 550 \cdot F \cdot (\tau_{ZP}^*)^a] \qquad (5.43)$$

The so-called flow parameter F is a function of the *Reynolds* number of the condensate as given by Equation (5.27), the ratio of the density and viscosity of the phases.

$$F = \frac{\max\left[(2 \cdot Re_l)^{0.5}; \, 0.132 \cdot Re_l^{0.9}\right]}{Re_g^{0.9}} \cdot \frac{\eta_l}{\eta_g} \cdot \sqrt{\frac{\rho_g}{\rho_l}} \qquad (5.44)$$

The exponent a takes into account the relation between sheer stress and gravity force.

$$a = \begin{cases} 0.30 & \text{for} & \tau_g^* \leq 1 \\ 0.85 & \text{for} & \tau_g^* > 1 \end{cases} \tag{5.45}$$

The correction terms C_{lam} and C_{turb} are defined as:

$$C_{lam} = 1 + (Pr_l^{0.56} - 1) \cdot \tanh(\tau_{ZP}^*) \qquad C_{turb} = 1 + (Pr_l^{0.08} - 1) \cdot \tanh(\tau_{ZP}^*) \tag{5.46}$$

The film thickness δ^+ is also a function of the flow parameter F.

$$\frac{\delta^+}{d} = \frac{6.59 \cdot F}{\sqrt{1 + 1400 \cdot F}} \tag{5.47}$$

The dimensionless sheer stress as defined in Equation (5.43) must de determined by iteration.

The recalculation of an existing heat exchanger or the design of a new one require different procedures. In both cases the calculation has to be done in several steps.

At design calculation, the tube diameter, the mass flow rate of the produced condensate and the heat transfer coefficient outside the tube must be known. Nodes are selected on which the local heat transfer coefficients are calculated and its mean value is the one between two nodes. The nodes represent vapor qualities. At the start of condensation the mass flow rate of condensate is 0 and the heat transfer coefficient infinite. Therefore, for the first node usually a very small condensate flow rate, e.g. 1 % ($x = 0.99$), is selected. The heat transfer coefficients are determined with mean vapor quality between the nodes (e.g. between nodes $x = 1$ and $x = 0.99$ calculation with $x_m = 0.995$). The required tube length for the production of the condensate between two nodes is calculated. The required total tube length is the sum of the calculated partial tube lengths.

For the recalculation of an existing heat exchanger with an assumed condensate mass flow rate, the same procedure is performed as for the design. The condensate flow rate has to be varied as long as the given tube length is reached.

5.2.1.2   Counterflow (vapor flow upward)

The calculation is similar to that for parallel flow, but the dimensionless sheer stress has a much higher value and is defined as:

$$\tau_{ZP}^* = \tau_g^* \cdot [1 + 1400 \cdot (\tau_{ZP}^*)^a] \tag{5.48}$$

At high vapor velocities the condensate can clog the tube and an oscillating flow will occur, which normally does not allow continuous operation and must therefore be avoided. The condenser must designed so that no clogging can occur. There exists a critical *Weber* number of $We_c = 0.01$, which should not be exceeded. The *Weber* number is defined as:

$$We = \frac{\tau_{ZP} \cdot \delta^+}{\sigma_l} = \frac{\tau_{ZP}^* \cdot \rho_l \cdot g \cdot (\delta^+)^2}{\sigma_l} \tag{5.49}$$

The two phase flow sheer stress τ_{ZP}^* is given by Equation (5.43), δ^+ with Equation (5.47), σ_l is the surface tension.

EXAMPLE 5.5: Design of the condenser of Example 5.3 with condensation in the tubes

A condenser with the parameters as given in Example 5.3 is to be designed. The number of tubes is 47. The heat transfer coefficient outside the tube is 6500 W/(m² K), the dynamic viscosity of the vapor $\eta_g = 14.2 \cdot 10^{-6}$ kg/(m s).

Find

The required tube length.

Solution

Schematic   Temperature profile as in Example 5.3

Assumption

- The mean outside heat transfer coefficient is constant.

Analysis

The local heat transfer coefficients in the tubes are determined with Equation (5.39). For the calculation five nodes are selected with: $x = 0.99, 0.75, 0.5, 0.25, 0.0$. The local heat transfer coefficients are calculated with the mean vapor qualities between the nodes. The calculated parameters are given as functions of the vapor quality x. In the following formula the numeric values determined with mean vapor quality between the first and second node ($x_1 = 1, x_2 = 0.99$ $x_m = 0.995$) are given. Later on the calculated parameters for the other nodes are tabulated.

$$Re_l(x) = \frac{\Gamma}{\eta_l} = \frac{\dot{m}_{R134a} \cdot (1-x)}{n \cdot \pi \cdot d_i \cdot \eta_l} = \frac{0.5 \cdot (1-x)}{47 \cdot \pi \cdot 0.01 \cdot 142.7 \cdot 10^{-6}} = 10.522$$

$$Re_g(x) = \frac{c_g \cdot d_i}{v_g} = \frac{4 \cdot \dot{m}_{R134a} \cdot x \cdot d_i}{\eta_g \cdot n \cdot \pi \cdot d_i^2} = 88531$$

$$\xi(x) = 0.184 \cdot Re_g(x)^{-0.2} = 0.019$$

$$\tau_g(x) = \frac{\xi \cdot \rho_g \cdot c_g^2}{8} = \frac{2 \cdot \xi(x) \cdot \dot{m}_{R134a}^2 \cdot x^2}{n^2 \cdot \pi^2 \cdot d_i^4 \cdot \rho_g} = 0.508 \text{ Pa}$$

$$F(x) = \frac{\max\left(\sqrt{2 \cdot Re_l(x)}; 0.132 \cdot Re_l(x)^{0.9}\right)}{Re_g(x)^{0.9}} \cdot \frac{\eta_l}{\eta_g} \cdot \sqrt{\frac{\rho_g}{\rho_l}} = 4.197 \cdot 10^{-4}$$

$$\delta(x) = \frac{6.59 \cdot d_i \cdot F(x)}{\sqrt{1 + 1400 \cdot F(x)}} = 2.195 \cdot 10^{-5} \text{ m}$$

$$\tau_g^*(x) = \frac{\tau_g(x)}{g \cdot \rho_l \cdot \delta(x)} = 2.140$$

The dimensionless two-phase flow sheer stress is determined with Equation (5.43) by iteration.

$$\tau_{ZP}^*(x) = \tau_g^*(x) \cdot [1 + 550 \cdot F(x) \cdot (\tau_{ZP}^*(x))^a] = 3.612$$

$$C_{lam}(x) = 1 + (Pr_l^{0.56} - 1) \cdot \tanh(\tau_{ZP}^*(x) = 1.897$$

$$C_{turb}(x) = 1 + (Pr_l^{0.08} - 1) \cdot \tanh(\tau_{ZP}^*(x) = 1.096$$

The local *Nußelt* numbers of the laminar and turbulent condensate film are determined with Equations (5.30) and (5.34).

$$Nu_{L', lam, x} = \frac{\alpha_x \cdot L'}{\lambda_l} = 0.693 \cdot \left(\frac{1 - \rho_g / \rho_l}{Re_l}\right)^{1/3} \cdot f_{well} = 0.340$$

$$Nu_{L', turb, x} = \frac{\alpha_x \cdot L'}{\lambda_l} = \frac{0.0283 \cdot Re_l^{7/24} \cdot Pr_l^{1/3}}{1 + 9.66 \cdot Re_l^{-3/8} Pr_l^{-1/6}} = 0.0191$$

The index x represents the length. The *Nußelt* numbers are also a function of the vapor quality x and we receive:

$$Nu_{L', x}^* = (1 + \tau_{ZP}^*)^{1/3} \cdot \sqrt{(C_{lam} \cdot Nu_{L', lam, x})^2 + (C_{turb} \cdot Nu_{L', turb, x})^2} = 1.075$$

In this tube section the local and overall heat transfer coefficients are given as:

$$\alpha_x = \frac{Nu_{L', x}^* \cdot \lambda_l}{L'} = Nu_{L', x}^* \cdot \lambda_l \cdot \sqrt[3]{g / \nu^2} = 6\,302 \quad \frac{\text{W}}{\text{m}^2 \cdot \text{K}}$$

$$k = \left(\frac{1}{\alpha_a} + \frac{d_a}{2 \cdot \lambda_R} \cdot \ln\left(\frac{d_a}{d_i} \right) + \frac{d_a}{d_i \cdot \alpha_x} \right) = 2880 \ \frac{W}{m^2 \cdot K}$$

In this first section the vapor quality x is reduced from 1 to 0.99, i.e. 1 % of the vapor is condensed. The heat rate in this section is:

$$\dot{Q} = \dot{m}_{R134a} \cdot (x_0 - x_1) \cdot r = 795 \ W$$

The temperature of the cooling water can be given as a function of the vapor quality.

$$\vartheta_1(x) = \vartheta_1' + \frac{\dot{m}_{R134a} \cdot (1-x)}{\dot{m}_{KW} \cdot c_{pKW}}$$

The mean log temperature difference in the section x_1 to x_2 is:

$$\Delta\vartheta_m(x_1, x_2) = \frac{\vartheta_1(x_1) - \vartheta_1(x_2)}{\ln \dfrac{\vartheta_2 - \vartheta_1(x_2)}{\vartheta_2 - \vartheta_1(x_2)}}$$

In the first section the vapor quality changes from 1.0 to 0.99. The log mean temperature in this section is 5.025 K. The surface area, respectively the tube length required for the transfer of the heat rate, can be determined with overall heat transfer coefficient and the log mean temperature.

$$\Delta l(\Delta x) = \frac{\Delta A(\Delta x)}{\pi \cdot n \cdot d_a} = \frac{\dot{Q}(\Delta x)}{k(x) \cdot \Delta\vartheta_m \cdot \pi \cdot n \cdot d_a} = 0.026 \ m$$

In the table below the numeric values of the calculations are presented.

x	x_m	Rel	Re_g	τ_g^* 10^{-3}	τ_{ZP}^*	$Nu_{L,lam}$	$Nu_{L,turb}$	Nu_L	Q W	$\Delta\vartheta_m$ °C	k W/m² K	Δl m	l m
1.00													0
	0.995	11	88531	2.140	3.612	0.340	0.019	1.075	759	5.025	2880	0.026	
0.99													0.026
	0.870	274	77409	0.466	1.157	0.131	0.108	0.330	18216	5.629	1286	1.259	
0.75													1.285
	0.625	789	55610	0.126	0.688	0.096	0.175	0.282	18975	6.856	1133	1.223	
0.50													2.508
	0.375	1315	33366	0.031	0.302	0.083	0.219	0.271	18975	8.109	1095	1.070	
0.25													3.578
	0.125	1841	11122	0.002	0.041	0.075	0.252	0.268	18975	9.361	1085	0.935	
0.00													**4.513**

Discussion

The heat transfer in the tubes is increased by the vapor flow and therefore the required tube length is reduced. The reduction in this example with 1.5 m is rather considerable. With high vapor velocities the heat transfer coefficient could even be doubled, this would require – or allow – shorter tubes.

5.3.2   Condensation in horizontal tubes

In very short horizontal tubes in which the vapor velocity can be neglected the *Nußelt* number can be calculated as described in Chapter 5.1.1.5 with Equation (5.23) wherein for the wall length the inner diameter of the tube has to inserted. In longer tubes the mass flow rate of vapor increases and therefore the vapor inlet velocity. With increasing vapor velocity the contribution of gravity force decreases and the sheer stress determines the heat transfer coefficients. For the condensation in horizontal tubes the *Nußelt* number in Equation (5.39) is modified to:

$$Nu^*_{L',x} = \tau_g^{*1/3} \cdot \sqrt{(C_{lam} \cdot Nu_{L',lam,x})^2 + (C_{turb} \cdot Nu_{L',turb,x})^2} \qquad (5.50)$$

The dimensionless sheer force is determined with the vapor velocity c_g, which is calculated with the vapor volume ratio ε as a function of the flow parameter F according to Equation (5.44).

$$\varepsilon = 1 - \frac{1}{1 + \dfrac{1}{8.48 \cdot F}} \qquad (5.51)$$

With the vapor volume rate the condensate film thickness can be determined.

$$\delta = 0.25 \cdot (1 - \varepsilon) \cdot d \qquad (5.52)$$

The vapor velocity c_g is calculated with the vapor flow area inside the condensate film.

$$c_g = \frac{4 \cdot \dot{m} \cdot x}{\rho_g \cdot \pi \cdot (d_i - 2 \cdot \delta)^2} \qquad (5.53)$$

The sheer stress is determined with Equations (5.40) to (5.42):

$$\tau_g = \frac{0.184 \cdot Re_g^{-0.2}}{8} \cdot c_g^2 \cdot \rho_g \qquad (5.54)$$

The dimensionless sheer stress is determined with the film thickness as:

$$\tau_g^* = \frac{\tau_g}{g \cdot \rho_l \cdot \delta} \cdot (1 + 850 \cdot F) \qquad (5.55)$$

To determine the *Nußelt* number the procedure as described in Chapter 5.3.1.1 is applied.

EXAMPLE 5.6: Design of a refrigerator condenser

In the condenser of a refrigerator the heat rate of 1.0 kW shall be transferred to the air. Freon R134a is the refrigerant. The condensation temperature is 50 °C. The outside heat transfer coefficient related to the outer diameter of the finned tubes was determined as 400 W/(m² K). The ambient air temperature is 22 °C. The condenser tube outer diameter is 8 mm, its heat conductivity 372 W/(m² K) and the wall thickness 1 mm. The sketch below shows the arrangement of the condenser. The tube bends can be neglected. Material properties of Freon 134a:

$\rho_l = 1\ 102$ kg/m³, $\rho_g = 66.3$ kg/m³, $\lambda_l = 0.071$ W/(m K), $\eta_l = 142.7 \cdot 10^{-6}$ kg/(m s), $\eta_g = 13.5 \cdot 10^{-6}$ kg/(m s), $Pr_l = 3.14$, $r = 151.8$ kJ/kg.

Find

The required length of the tube.

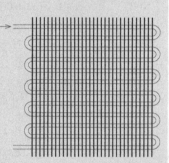

Solution

Schematic   See sketch.

Assumptions

- The mean heat transfer coefficients are constant.
- The tube bends can be neglected.
- The change in pressure is neglected.
- The temperature rise of the air is neglected.

Analysis

As in the vertical tubes the calculation is performed again in sections. The selected nodes are: $x = 1$, 0.75, 0.5, 0.25 and 0.0. For the calculation the following mean vapor qualities are used: 0.875, 0.625, 0.375 and 0.125. In the following the numeric values of in the first section with the vapor quality of $x = 0.875$ are shown. The other values are tabulated later.

The mass flow rate of the Freon, which condenses completely, can be determined with the given heat rate.

$$\dot{m} = \frac{\dot{Q}}{r} = \frac{1.0 \text{ kW}}{151.8 \text{ kJ/kg}} = 6.588 \cdot 10^{-3} \text{ kg/s}$$

The calculation of the *Nußelt* number follows here without comments.

$$Re_l(x) = \frac{\Gamma}{\eta_l} = \frac{\dot{m} \cdot (1-x)}{\pi \cdot d_i \cdot \eta_l} = \frac{6.588 \cdot 10^{-3} \cdot (1-x)}{\pi \cdot 0.006 \cdot 142.7 \cdot 10^{-6}} = 306.1$$

$$Re_g(x) = \frac{c_g \cdot d_i}{v_g} = \frac{4 \cdot \dot{m} \cdot x \cdot d_i}{\eta_g \cdot \pi \cdot d_i^2} = 90\,607$$

$$\xi(x) = 0.184 \cdot Re_g(x)^{-0.2} = 0.0190$$

$$F(x) = \frac{\max\left(\sqrt{2 \cdot Re_l(x)};\, 0.132 \cdot Re_l(x)^{0.9}\right)}{Re_g(x)^{0.9}} \cdot \frac{\eta_l}{\eta_g} \cdot \sqrt{\frac{\rho_g}{\rho_l}} = 4.946 \cdot 10^{-3}$$

$$\varepsilon(x) = 1 - \frac{1}{1 + 1/[8.48 \cdot F(x)]} = 0.9596$$

$$\delta(x) = 0.25 \cdot (1-\varepsilon) \cdot d_i = 6.034 \cdot 10^{-5} \text{ m}$$

$$c_g(x) = \frac{4 \cdot \dot{m} \cdot x}{\pi \cdot \rho_g \cdot (d_i - 2 \cdot \delta)^2} = 3.202 \text{ m/s}$$

$$\tau_g(x) = \frac{\xi \cdot \rho_g \cdot c_g^2}{8} = 2.446 \text{ Pa}$$

$$\tau_g^*(x) = \frac{\tau_g(x)}{g \cdot \rho_l \cdot \delta(x)} \cdot \left[1 + 850 \cdot F(x)\right] = 12.722$$

$$C_{lam}(x) = 1 + (Pr_l^{0.56} - 1) \cdot \tanh(\tau_{ZP}^*(x))] = 1.898$$

$$C_{turb}(x) = 1 + (Pr_l^{0.08} - 1) \cdot \tanh(\tau_{ZP}^*(x))] = 1.096$$

The local *Nußelt* numbers are determined with Equations (5.30) and (5.34).

$$Nu_{L', lam, x} = \frac{\alpha_x \cdot L'}{\lambda_1} = 0.693 \cdot \left(\frac{1 - \rho_g / \rho_l}{Re_l}\right)^{1/3} \cdot f_{well} = 0.127$$

$$Nu_{L', turb, x} = \frac{\alpha_x \cdot L'}{\lambda_1} = \frac{0.0283 \cdot Re_l^{7/24} \cdot Pr_l^{1/3}}{1 + 9.66 \cdot Re_l^{-3/8} Pr_l^{-1/6}} = 0.1138$$

Here the index x indicates the length of the section for which the local *Nußelt* numbers were determined. The *Nußelt* number is a function of the vapor quality x.

$$Nu_{L', x}^* = \tau_g^{*1/3} \cdot \sqrt{(C_{lam} \cdot Nu_{L', lam, x})^2 + (C_{turb} \cdot Nu_{L', turb, x})^2} = 0.632$$

The characteristic length L is:

$$L' = \sqrt[3]{v_l^2 / g} = 1.571 \cdot 10^{-5} \text{ m}$$

The local and the overall heat transfer coefficients in the first section are:

$$\alpha_x = \frac{Nu_{L',x}^* \cdot \lambda_l}{L'} = 2443 \ \frac{\text{W}}{\text{m}^2 \cdot \text{K}}$$

$$k = \left(\frac{1}{\alpha_a} + \frac{d_a}{2 \cdot \lambda_R} \cdot \ln\left(\frac{d_a}{d_i} \right) + \frac{d_a}{d_i \cdot \alpha_x} \right) = 328 \ \frac{\text{W}}{\text{m}^2 \cdot \text{K}}$$

In each section 25 % of the entering vapor is condensed, i.e. the heat rate in all sections has the value:

$$\Delta \dot{Q} = \dot{m} \cdot (x_0 - x_1) \cdot r = 250 \text{ W}$$

To be able to transfer the given heat rate a certain heat exchanger surface area, i.e. tube length, must be determined.

$$\Delta l = \frac{\Delta A}{\pi \cdot d_a} = \frac{\Delta \dot{Q}}{\pi \cdot d_a \cdot k \cdot (\vartheta_F - \vartheta_L)} =$$

$$= \frac{250 \cdot \text{W} \cdot \text{m}^2 \cdot \text{K}}{\pi \cdot 0,008 \cdot \text{m} \cdot 328 \cdot \text{W} \cdot (50 - 22) \cdot \text{K}} = 1.055 \text{ m}$$

The table below shows the calculated values.

x_m	τ_g	τ_{ZP}^*	α_i	k	ΔL
-	Pa	-	W/(m² K)	W/(m² K)	m
0.875	1.593	12.722	2 857	336.7	1.055
0.625	1.064	8.099	2 443	328.8	1.083
0.375	0.604	5.419	2 350	325.7	1.091
0.125	0.200	3.350	2 185	321.2	1.106

The total tube length is the sum of the section tube lengths, i.e. **4.335 m**.

Discussion

The outside heat transfer coefficient is reasonably lower than that in the tubes. For the air side a mean heat transfer was given and the air temperature was assumed as constant. To take the temperature rise of the air into account, knowledge of a cross flow heat exchanger is required, which will be discussed in Chapter 8. However, this example is rather close to reality where about 5 to 10 % larger surface area would be determined.

EXAMPLE 5.7: Condenser retrofit

The condenser of a U.S. power station had an insufficient performance, i.e. the condenser pressure was higher than its designed value. It was planned to replace condenser bundles with state of the art bundles. In the call for tenders the following features were given:

Each of the three low-pressure turbines has a condenser, in which the water is passing in serial flow (see sketch). The heat rate to each condenser is 733 MW.

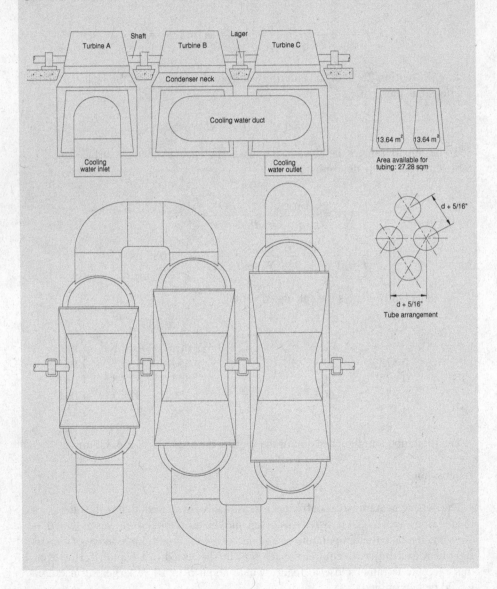

Data of the existing condenser:

Number of tubes in each condenser: 27000 tubes with 1" OD and 1.2 mm wall thickness

Distance between tubes: $d_a = 1\ 5/16"$

Tube lengths: $l_A = 35'$ $l_B = 45'$ $l_C = 55'$

Cooling water volume flow rate: 381 500 GPM

                         at 86 ft water column pressure head

In the specification it was required that with the new condensers at a cooling water inlet temperature of 35 °C to condenser A, the pressure in condenser C does not exceed 5 inHg column. The tube lengths and the outer shell dimensions must be kept. The tube material requested was titanium with 0.7 mm wall thickness.

The successful bidder had a state of the art bundle design, which required within the given shell dimensions for the tubing a surface area of 27.28 m². The optimization for the tube outer diameter was done from 1" in 1/8" steps. A gap of 5/16" between the tubes was selected.

In the above sketch the disposition of the condensers is shown left, the area for the tubing and the tube arrangement is presented on the right.

For the cooling water the following material properties can be used:

$c_p = 4.174$ kJ/(kg K), $\rho = 990.2$ kg/m³, $\nu = 0.6 \cdot 10^{-6}$ m²/s, $\lambda = 0.6368$ W/(m K), $Pr = 3.946$.

Conversion coefficients: 1 GPM (gallon per minute) = 0.063083 l/s, in = 0.0254 m, ft = 0.3048 m, 1 inHg = 3 386.39 Pa.

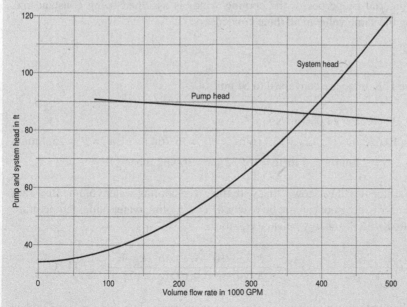

Pressure-flow characteristic of the cooling water pump and cooling water system

The cooling water flow rate is a function of the pressure drop in the condenser according to the cooling water pump characteristic. The pressure drop in the condenser tubes can be calculated with the formula of *Blasius*:

$$\Delta p_v = 0.3164 \cdot Re^{-0.25} \cdot l \,/\, d \cdot (c^2 \cdot \rho \,/\, 2)$$

Find

The design with which the condenser C pressure is below 5 inHg at 35 °C cooling water temperature and the pressure in the three condensers.

Advice: First check the outlet temperature of the cooling water.

Solution

Schematic   See sketch in the definition of the example.

Assumptions

- In the ducts to and from the condenser tubes the pressure drop is proportional to the square of the volume flow rate.
- The static pressure head is always existent (height difference between cooling water level and injection nozzles in the cooling tower).
- The vapor pressure drop in the condenser bundles is can be neglected.
- The direction of the heat flux and of the tube length can be neglected.
- The material properties of the cooling water is assumed being constant and having the same value in all three condensers.

Analysis

First the U.S. units are conversed to SI units.

$\dot{V}_0 = 381500$ GPM $= 24.066$ m^3/s

$l_A = 35' = 10.668$ m $l_B = 13.716$ m $l_C = 16.764$ m $d_a = 1'' = 25.4$ mm

$d_i = d_a - 2 \cdot s = 23$ mm   5 inHg $= 16932$ Pa $= 169.32$ mbar

As a first step it will be shown, that with the given cooling water volume flow rate the condenser C pressure can not be reached. The cooling water outlet temperature is calculated with the energy balance equation.

$$\vartheta_1'' = \vartheta_1' + \frac{\dot{Q}_{tot}}{\dot{V}_0 \cdot \rho_1 \cdot c_{p1}} = 35 \ ^\circ\mathrm{C} + \frac{3 \cdot 733 \cdot 10^6 \cdot \mathrm{W \cdot s \cdot m^3 \cdot kg \cdot K}}{24.066 \ \mathrm{m^3} \cdot 990.2 \cdot \mathrm{kg} \cdot 4174 \cdot \mathrm{J}} = 57.11 \ ^\circ\mathrm{C}$$

According to the steam table the saturation temperature at 5 inHg = 169.32 mbar is 56.5 °C, i.e. the cooling water outlet temperature is higher than the saturation temperature of the vapor, which should heat up the cooling water. To reach the desired condenser pressure of 5 inHg, the cooling water must be at least 2 to 3 K colder than the saturation temperature of the vapor. This can only be reached with a higher cooling water flow rate. The most economic solution would be to reduce the pressure drop in the condenser tubes so that the cooling water volume flow rate increases according to the cooling water pump characteristics. The pressure head of the pump is the sum of the constant static Δp_{st} plus the dynamic head of the cooling water system, which is the frictional loss Δp_{va} in the cooling water ducts and the frictional pressure drop in the condenser tubes Δp_v. The latter one can be influenced by the design of the new bundles. The pressure drop in the ducts and water boxes will not change for the new condensers and their value can be determined with the given data of the old condenser. The pressure drop in the condenser tubes of the old condenser is:

$$\Delta p_{v0} = 0.3164 \cdot Re_{d_i}^{-0.25} \cdot \frac{l_{ges}}{d_i} \cdot \frac{c_0^2 \cdot \rho_1}{2}$$

For the velocity in the tubes we receive:

$$c_0 = \frac{4 \cdot \dot{V}_0}{n \cdot \pi \cdot d_i^2} = \frac{4 \cdot 24.066 \text{ m}^3}{27\,000 \cdot \pi \cdot 0.023^2 \cdot \text{m}^2 \cdot \text{s}} = 2.143 \frac{\text{m}}{\text{s}}$$

The *Reynolds* number is: $Re_{d_i} = \dfrac{c_0 \cdot d_i}{v_1} = \dfrac{2.145 \cdot 0.023}{0.6 \cdot 10^{-6}} = 82\,239$

The frictional pressure drop calculated with *Blasius* formula:

$$\Delta p_{v0} = 0.3164 \cdot 82\,239^{-0.25} \cdot \frac{41.148 \cdot \text{m}}{0.023 \cdot \text{m}} \cdot \frac{2.145^2 \cdot \text{m}^2 \cdot 990.2 \cdot \text{kg}}{2 \cdot \text{s}^2 \cdot \text{m}^3} =$$
$$= 76169 \text{ Pa} = 25.735 \text{ ft H}_2\text{O}$$

The pump head at 381500 GPM volume flow rate is 86 ft. Therein is 34 ft the static head and 25.74 ft the frictional pressure drop in the condenser tubes. For the pressure drop in the duct to and from the condenser remain 26.27 ft. The system characteristic with new condenser bundles can be calculated. It changes with the number and size of the tubes. The frictional pressure drop Δp in the new condenser tubes can be given as a ratio to the pressure drop with old condenser Δp_{v0}:

$$\frac{\Delta p_v}{\Delta p_{v0}} = \left(\frac{Re}{Re_0} \right)^{-0.25} \cdot \frac{d_{i0}}{d_i} \cdot \frac{c^2}{c_0^2}$$

The tube velocity can be calculated as a function of volume flow rate, tube number, tube inner diameter and inserted in the above equation.

$$\Delta p_v = \Delta p_{v0} \cdot \left(\frac{n_0 \cdot d_{i0}^2 \cdot \dot{V} \cdot d_i}{n \cdot d_i^2 \cdot \dot{V}_0 \cdot d_{i0}} \right)^{-0.25} \cdot \frac{d_{i0}}{d_i} \cdot \left(\frac{n_0 \cdot d_{i0}^2 \cdot \dot{V}}{n \cdot d_i^2 \cdot \dot{V}_0} \right)^2 = 25.74 \cdot \text{ft} \cdot \left(\frac{n_0 \cdot \dot{V}}{n \cdot \dot{V}_0} \right)^{1.75} \cdot \left(\frac{d_{i0}}{d_i} \right)^{4.75}$$

The system head is the sum of the static head and the dynamic heads, caused by friction.

$$\Delta p_{tot} = \left[34 + 26.27 \cdot \left(\frac{\dot{V}}{\dot{V}_0} \right)^2 + 25.74 \cdot \left(\frac{n_0 \cdot \dot{V}}{n \cdot \dot{V}_0} \right)^{1.75} \cdot \left(\frac{d_i}{d_{i0}} \right)^{4.75} \right] \cdot \text{ft}$$

With this equation the system head for the new condensers can be calculated and with the pump characteristic the volume flow rate can be determined. For the new condensers first the number of tubes which can be fitted into the given surface of 27.28 m² has to determined for different tube diameters. For one tube in 60° arrangement the following area is required:

$$A_0(d_a) = \frac{\sqrt{3}}{2} \cdot (d_a + s_0)^2$$

For the number of tubes as a function of the outer diameter we receive:

$$n(d_a) = \frac{27.28 \cdot \text{m}^2}{\sqrt{3} \cdot (d_a + 0.0079375 \text{ m})^2}$$

For tubes with 1" outer diameter 28343 tubes can be inserted, with 1 1/8" 23628 and with 1 1/4" 20000.

The volume flow rate resulting from change of the system characteristic can be determined either by iteration or graphically in a diagram. The changed system diagram can be drawn in the diagram and the intersection point with the pump characteristic is the new volume flow rate. Below a detail of the system and pump characteristic diagram made with *Origin* soft ware is shown.

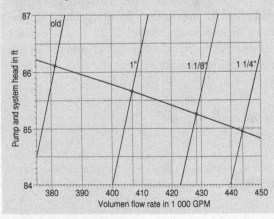

With the three investigated tube diameters the following volume flow rate were obtained:

1"	406800 GPM	25.662 m³/s
1 1/8"	428500 GPM	27.031 m³/s
1 1/4"	443900 GPM	28.002 m³/s

For the 1" tube diameter the cooling water outlet temperature is:

$$\vartheta_1'' = \vartheta_1' + \frac{\dot{Q}_{tot}}{\dot{V}_0 \cdot \rho_1 \cdot c_{p1}} = 35\ °C + \frac{3 \cdot 733 \cdot 10^6 \cdot W \cdot s \cdot m^3 \cdot kg \cdot K}{25.662\ m^3 \cdot 990.2 \cdot kg \cdot 4174 \cdot J} = 55.73\ °C$$

The increased flow rate is due to the larger inner diameter of the titanium tubes. The cooling water outlet temperature is 1.9 K lower than the saturation temperature of the vapor at 5 in Hg saturation pressure. It is still too high to transfer the heat rate with the heat exchanging surface area.

The following outlet temperatures were calculated:
1 1/8" tubes	54.68 °C
1/4" tubes	54.00 °C

With the 1.25" tube diameter the cooling water outlet temperature is 2.5 K lower than the saturation temperature. With this value the condenser can most probably be designed, but we have to check it.

With the equation given in Chapter 3 first the heat transfer coefficients in the tubes can be calculated. The velocity in the tubes is:

$$c = \frac{4 \cdot \dot{V}}{n \cdot \pi \cdot d_i^2} = \frac{4 \cdot 28.002 \cdot m^3 / s}{20\,000 \cdot \pi \cdot 0.03035^2 \cdot m^2} = 1.935\ \frac{m}{s}$$

For the *Reynolds* number we receive:

$$Re_{d_i} = c \cdot d_i / \nu = 1.935 \cdot 0.03035 / 0.6 \cdot 10^{-6} = 97897$$

The hydraulic resistance coefficient required for the determination of the *Nußelt* number is:

$$\xi = \left[1.8 \cdot \lg(Re_{d_i}) - 1.5\right]^{-2} = 0.018$$

Nußelt number:

$$Nu_{d_i} = \frac{(\xi/8) \cdot Re_{d_i} \cdot Pr_1}{1 + 12.7 \cdot \sqrt{\xi/8} \cdot (Pr^{2/3} - 1)} = \frac{(0.018/8) \cdot 97\,897 \cdot 3.946}{1 + 12.7 \cdot \sqrt{0.018/8} \cdot (3.946^{2/3} - 1)} = 454.2$$

Heat transfer coefficient in the tubes:

$$\alpha_i = Nu_{d_i} \cdot \lambda_1 / d_i = 9530.7 \ W / (m^2 \cdot K)$$

The thermal resistance in the tubes and in the tube walls results as:

$$R_i = \frac{d_a}{d_i \cdot \alpha_i} + \frac{d_a}{2 \cdot \lambda_R} \cdot \ln\left(\frac{d_a}{d_i}\right) = \frac{31.75}{30.35 \cdot 9530.7} + \frac{0.03175}{2 \cdot 15} \cdot \ln\left(\frac{31.75}{30.35}\right) = 0.15749 \ \frac{m^2 \cdot K}{kW}$$

For the condensation the determination of the heat transfer coefficients must be performed separately for each condenser. As the saturation pressures are not known, we have to start with an estimated value.

Condenser A:

It is assumed that the saturation temperature is 4 K higher than the cooling water outlet temperature of condenser A. As the heat rate in all three condensers has the same magnitude and as specific heat capacity of the water is assumed to be constant, the temperature rise in each condenser is 6.33 K, i.e. the condenser A cooling water outlet temperature is 41.33 °C and the saturation temperature 45.33 °C. In condenser A we have a pressure of 97.60 mbar. The mean temperature of the condensate is estimated with 43 °C. From the vapor table we receive the following values:

$\rho_l = 991 \ kg/m^3$, $\rho_g = 0.0666 \ kg/m^3$, $\eta_l = 617.8 \cdot 10^{-6} \ kg/(m \ s)$, $v_l = 0.623 \cdot 10^{-6} \ m^2/s$, $\lambda_l = 0.6347 \ W/(m \ K)$, $r = 2393.2 \ kJ/kg$.

The mass flow rate per unit width is determined with Equation (5.28):

$$\Gamma = \frac{\dot{Q}}{r \cdot n \cdot l_A} = \frac{733000 \cdot kW \cdot kg}{2393.2 \cdot kJ \cdot 20000 \cdot 10.668 \cdot m} = 1.436 \cdot 10^{-3} \ \frac{kg}{m \cdot s}$$

Reynolds number: $Re_l = \Gamma / \eta_l = 1.436 \cdot 10^{-3} / 0.6178 \cdot 10^{-3} = 2.325$

Nußelt number: $Nu_L = 0.959 \cdot \left(\frac{1 - \rho_g / \rho_l}{Re_l}\right)^{1/3} = 0.724$

Characteristic length:

$$L = \sqrt[3]{v_l^2 / g} = \sqrt[3]{(0.623 \cdot 10^{-6})^2 / 9.806} = 3.408 \cdot 10^{-5} \ m$$

Outside heat transfer coefficient:

$$\alpha = Nu_L \cdot \lambda_1 / L = 0.724 \cdot 0.6374 / 3.408 \cdot 10^{-5} = 13\,483 \ W/(m^2 \cdot K)$$

Now the overall heat transfer coefficient, the log mean temperature difference, the saturation temperature and pressure of condenser A can be determined.

$$k = \left(\frac{1}{\alpha} + R_i\right)^{-1} = \left(\frac{1}{13483} + 0.1575 \cdot 10^{-3}\right)^{-1} = 4317 \ \frac{W}{m^2 \cdot K}$$

$$\Delta\vartheta_m = \frac{\dot{Q}}{k \cdot n \cdot \pi \cdot d_a \cdot l_a} = \frac{733 \cdot MW \cdot m^2 \cdot K}{4317 \cdot W \cdot 20000 \cdot \pi \cdot 0.03175 \cdot m \cdot 10.668 \cdot m} = 7.979 \ K$$

$$\frac{\vartheta_{sA} - \vartheta'_A}{\vartheta_{sA} - \vartheta''_A} = \Theta = \exp\left(\frac{\vartheta'_A - \vartheta''_A}{\Delta\vartheta_m}\right) = 2.212$$

$$\vartheta_{sA} = \frac{\vartheta'_A - \vartheta''_A \cdot \Theta}{1 - \Theta} = \frac{35 \ °C - 2.212 \cdot 41.41 \ °C}{1 - 2.212} = 46.56 \ °C$$

The saturation temperature is 1.2 K larger than the assumed value of 43 °C. With the calculated values the mean temperature of the condensate is:

$$\vartheta_m = \vartheta_{sA} - 0.5 \cdot (\vartheta_{WA} - \vartheta_{sA}) = \vartheta_{sA} - 0.5 \cdot \Delta\vartheta_m \cdot k / \alpha = 45.28 \ °C$$

In the condenser A we have a pressure of 103.91 mbar. The material properties of the condensate:

$\rho_l = 990.1$ kg/m^3, $\rho_g = 0.0707$ kg/m^3, $\eta_l = 593.1 \cdot 10^{-6}$ kg/(m s), $r = 2390.2$ kJ/kg, $\nu_l = 0.599 \cdot 10^{-6}$ m^2/s, $\lambda_l = 0.6377$ W/(m K).

With these material properties we receive a condenser pressure of **103.70 mbar** and a saturation temperature of **46.52 °C**, this is 0.04 K lower than the value calculated before. With this accuracy the iteration can be terminated.

Condenser B:

The procedure is the same as for condenser A. Here only the material properties and results are listed.

$\rho_l = 987.9$ kg/m^3, $\rho_g = 0.0889$ kg/m^3, $\eta_l = 545.0 \cdot 10^{-6}$ kg/(m s), $r = 2378.5$ kJ/kg, $\nu_l = 0.552 \cdot 10^{-6}$ m^2/s, $\lambda_l = 0.6438$ W/(m K).

The saturation temperature is **51.14 °C** and the pressure **130.67 mbar**.

Condenser C:

In analogy here we receive:

$\rho_l = 985.6$ kg/m^3, $\rho_g = 0.1115$ kg/m^3, $\eta_l = 502.8 \cdot 10^{-6}$ kg/(m s), $r = 2366.4$ kJ/kg, $\nu_l = 0.510 \cdot 10^{-6}$ m^2/s, $\lambda_l = 0.6494$ W/(m K).

Saturation temperature: **56.44 °C**, pressure: **168.80 mbar**.

The pressure of 168,8 mbar corresponds to the U.S. unit of 4.985 inHg which is lower than the requested 5 inHg.

Discussion

This example demonstrates that for the design and calculation of a heat exchanger with the science of heat transfer only, no result could be delivered. Here the determination of pressure drop, the use of pump characteristics and knowledge of the fluid mechanics and civil engineering were required.

According to the definition of the example, the influence of the tube length and the direction of the heat flux were not taken into account. If they were included approximately 3 to 5 % higher heat transfer coefficients would be calculated but the pressure drop in the bundle would create a reduction of approximately the same magnitude. Anyhow at the design of a heat exchanger oft this size a very exact calculation is required.

The condensers discussed here were designed by Brown Boveri & Cie, Switzerland. Two halves of each condensers were erected in a shop, transferred from Tulsa, Oklahoma by bark via the Panama channel to Oregon State where the power plant was located. In four weeks the existing condenser bundles were removed from the shell and the new ones inserted and welded to the shell. The price of the modules including transport and installation was about 18 Million US$. In the contact in case of a higher pressure than 5.285 inHg a penalty of 1.8 Mio. US$ per 0.1 in Hg was established. That means, a too small surface area which results in a too high pressure is punished with a penalty. A too large surface area however, results in higher costs and is therefore not competitive. The condenser discussed here had reached the guaranteed values.

6   Boiling heat transfer

For the design of heat exchangers in which steam or vapor is generated, knowledge of the laws of *boiling heat transfer* is required. Steam generators are used in steam power plants, heat pumps, refrigerators, boilers, distilling and rectifying columns. Boiling can occur in static and flowing fluids.

Evaporation occurs, when a liquid is heated up to saturation temperature ϑ_S and heat transfer to it is continued. When a small heat rate is transferred to a static saturated liquid, steam is released from it surface, its mass flow rate is determined by the heat rate. With increasing heat rate steam bubbles are generated at the heating surface and nucleate boiling starts.

Condensation starts independent of the steam temperature immediately when the steam has contact to a surface with a lower temperature than the saturation temperature. During boiling, however, it was found that on a surface with a temperature above saturation temperature, steam generation does not necessarily start. A fluid can be superheated without steam generation. In extreme cases, very high excess temperatures of more than 100 K were observed; this phenomenon is called *boiling retardation*. In such a case a sudden steam explosion can follow. The mechanism behind this phenomenon can be explained as follows: with a wall temperature above saturation temperature bubble generation starts. Because of the surface tension, the pressure in the bubble is larger than in the fluid and with the accordingly higher saturation temperature in the bubble, recondensation occurs. For the existence of the bubble the surrounding liquid must be superheated! Depending on the *excess temperature* of the wall $\Delta\vartheta = \vartheta_w - \vartheta_S$ and the velocity of the liquid, different types of heat transfer processes were observed.

6.1   Pool boiling

The process of heat transfer to a static liquid in a vessel with a heated surface on which steam bubbles are formed, is called *pool boiling*. In Figure 6.1 heat transfer coefficient and heat flux at pool boiling is shown versus the excess temperature of the wall.

With increasing excess temperature the formation of bubbles starts (B). This is the beginning of *nucleate boiling*. Bubbles are formed on the heating surface in small pits, called *nucleation sites*. With increasing excess temperature the intensity of bubble formation and the number of active nucleation sites increase. The bubble generation acts as a stirring of the liquid and increases the convective heat transfer in the liquid. The bubbles ascend to the surface of the liquid. As can be seen in Figure 6.1, the heat flux and heat transfer coefficients are increasing rapidly.

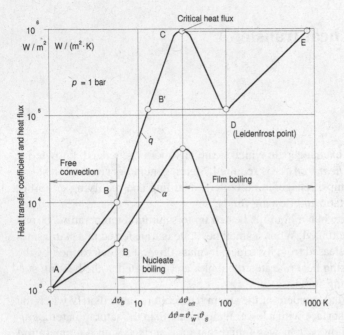

Figure 6.1: Heat transfer coefficient and heat flux of boiling of water vs. excess temperature

With further rising excess temperature and heat rate, at a certain stage the bubbles coalesce and form continuous steam film areas (*Leidenfrost phenomenon* - starting at point (C)). This process is called *film boiling*. The transition from nucleate to film boiling is called boiling crisis. In the vapor or steam film areas, the heat transfer is governed mainly by radiation and conduction. As the heat transfer resistance of the steam film areas is much larger than that of the boiling convection process, a further increase in heat flux beyond the critical heat flux requires a large increase of the wall temperature. This can happen in large steam generators where the heat is provided by combustion or nuclear fission at an almost constant heat flux. In such installations, the excess temperature jumps from point C to point E if the heat flux exceeds the critical value. With water at a pressure of 1 bar as shown in Figure 6.1, the jump of the excess temperature is 770 K. The heating wall temperature goes from 100 °C to 870 °C between C and E. Large steam generators are working at much higher pressures and the saturation temperature is correspondingly higher. Most materials for steam generators could not withstand such a large temperature change and a destruction of the boiler wall would be the result (burn out). Therefore, the transition from nucleate boiling to film boiling must be avoided. At the design of the boiler it must be made sure that the critical heat flux is never reached.

Point D can be reached when an excess temperature controlled heating process is maintained, where the excess temperature is increased slowly. D can also be reached, when dropping the heat flux coming from E. Then, at D a sudden decrease of the excess temperature to B happens and film boiling changes back to nucleate boiling. The physical state between point C and D can only be reached in laboratories using special liquids and devices.

Film boiling is a hazard for heat exchangers which should be avoided and its heat transfer coefficients, therfore, are not further discussed in this book.

The produced steam or vapor mass flow rate at boiling is calculated with the energy balance equation as:

$$\dot{m}_g = \dot{Q} / r \qquad (6.1)$$

This correlation is valid for all types of boiling.

6.1.1   Sub-cooled convection boiling

As long as no bubbles are generated in a static sub-cooled liquid, even if the heating surface temperature is larger than the saturation temperature, the heat transfer coefficients and the *Nußelt* numbers are calculated as described in Chapter 4. For horizontal tubes and horizontal plane surfaces simplified formula are proposed [6.1].

The *Nußelt* number is:

$$Nu_l = 0.15 \cdot (Gr_l \cdot Pr)^{1/3} \qquad (6.2)$$

The characteristic lengths for the *Grashof* and *Nußelt* number are defined as:

Square surface: $L = a \cdot b / 2\,(a + b)$
Circular surface: $L = d/4$
Horizontal cylinder: $L = d$

6.1.2   Nucleate boiling

As already mentioned at certain locations, the so called nucleation sites, bubbles are generated. The number of nucleation sites increases with increasing heat rate. The bubbles are growing from microscopic pits on the – technically rough – surface. The heat is first transferred to the boundary layer of the liquid and from there into the bubble. The pressure p_g in the bubble is larger than that in the liquid, due to the surface tension. Figure 6.2 shows the nascency of a steam bubble.

The overpressure in the steam bubble is generated by the force induced by the surface tension σ. The force balance equation of the pressure and surface tension forces is:

$$p_g - p_l = 4 \cdot \sigma / d \qquad (6.3)$$

For the nascency of a bubble with the diameter d a minimum excess temperature must exist.

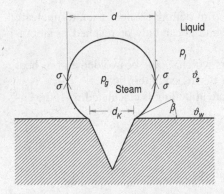

Figure 6.2: Nascency of a steam bubble

The diameter of the nucleation site d_K is that of the smallest bubble. According to the *Laplace-Kelvin*-derivation the correlation between the excess pressure and the excess temperature is:

$$\frac{p_g - p_l}{T_g - T_S} = \frac{\rho_g \cdot r}{T_s} \qquad (6.4)$$

With Equation (6.3) we receive for the required excess temperature:

$$\vartheta_W - \vartheta_S = \frac{4 \cdot \sigma \cdot T_s}{d_K \cdot \rho_g \cdot r} \qquad (6.5)$$

At a certain size the bubble tears off the surface and ascends to the liquid surface. It transfers the received heat as latent heat of evaporation. In the wake of the bubble a drift flow is generated, that increases the convective heat transfer. Equation (6.5) shows that with increasing excess temperature the smallest possible bubble diameter decreases and with it the number of nucleation sites.

For the determination of the heat transfer coefficients the rules of bubble nascency and its tear-off help to develop correlations. With the forces acting on a bubble a model for the *tear-off diameter* d_A of the bubble can be developed. With a great number of nucleation sites according to the frequency distribution of the uncoupling diameter the most probable diameter can be given. With this model and with experimental data the following tear-off diameter was found:

$$d_A = 0.0149 \cdot \beta^0 \cdot \sqrt{\frac{2 \cdot \sigma}{g \cdot (\rho_l - \rho_g)}} \qquad (6.6)$$

Therein β^0 is the *bubble contact angle*. Its value depends on the liquid. Below the bubble contact angels for a few liquids are listed.

Water: $\qquad\qquad\qquad\qquad$ $\beta^0 = 45°$

Refrigerants: $\qquad\qquad\qquad$ $\beta^0 = 35°$

Benzene: $\qquad\qquad\qquad$ $\beta^0 = 40°$

The heat transfer coefficient as a function of heat flux and excess temperature is:

$$\alpha_B = \frac{\dot{q}}{\vartheta_W - \vartheta_s} = \frac{\dot{q}}{\Delta\vartheta} \tag{6.7}$$

The characteristic length in the *Nußelt* number of nucleate boiling is the tear-off diameter.

$$Nu_{d_A} = \frac{\alpha_B \cdot d_A}{\lambda_l} \tag{6.8}$$

For the determination of the heat transfer coefficients reference values, either calculated or determined experimentally, are required.

For the heat transfer coefficient the following correlation was found [6.2]:

$$\alpha_B = \alpha_0 \cdot f(p*_0) \cdot \left(\frac{\lambda_l \cdot \rho_l \cdot c_{pl}}{\lambda_{l0} \cdot \rho_l 0 \cdot c_{pl0}}\right)^{0.25} \cdot \left(\frac{R_a}{R_{a0}}\right)^{0.133} \cdot \left(\frac{\dot{q}}{\dot{q}_0}\right)^{0.9-0.3 \cdot p*} \tag{6.9}$$

The reference heat transfer coefficient α_0 is determined at the dimensionless pressure of $p* = p/p_{krit} = 0.1$, R_a the *mean roughness index* according to DIN 4762/01.89, which takes into account the surface properties with regard to the number and size of nuclei. It replaces the *smoothing roughness* R_p according to DIN 4672/08.60 used earlier. Between the two values the following relationship exist:

$$R_a = 0.4 \cdot R_p \tag{6.10}$$

The formerly used reference value $R_{p0} = 1$ µm is replaced by $R_{a0} = 0,4$ µm.
The function $f(p*)$ takes into account the influence of pressure.

$$f(p*) = \begin{vmatrix} 1.73 \cdot p*^{0.27} + \left(6.1 + \dfrac{0.68}{1-p*}\right) \cdot p*^2 & \text{for water} \\[3mm] 1.2 \cdot p*^{0.27} + \left(2.5 + \dfrac{1}{1-p*}\right) \cdot p* & \text{for other pure liquids} \end{vmatrix} \tag{6.11}$$

Based on a large number of tests *Stephan* and *Preußer* [6.2] found at a pressure of $p = 0.03 \cdot p_{krit}$, heat flux of 20000 W/m² and a mean roughness index of $R_a = 0.4$ µm the following correlation for the reference *Nußelt* number Nu_{dA0} (Caution! Use the material properties at $p = p* \cdot 0.03$ for the calculation!):

$$Nu_{d_A 0} = 0.1 \cdot \left(\frac{\dot{q}_0 \cdot d_A}{\lambda_l \cdot T_s} \right)^{0.674} \cdot \left(\frac{\rho_g}{\rho_l} \right)^{0.156} \cdot \left(\frac{r \cdot d_A^2}{a_l^2} \right)^{0.371} \cdot \left(\frac{a_l^2 \cdot \rho_l}{\sigma \cdot d_A} \right)^{0.35} \cdot Pr_l^{-0.16} \quad (6.12)$$

Therein is $a_l = \lambda_l / (\rho_l c_{pl})$ the thermal diffusivity of the liquid.

$$\alpha_0 = \frac{f(0.1)}{f(0.03)} \cdot Nu_{d_A 0} \cdot \frac{\lambda_l}{d_A} = \frac{1}{f(0.03)} \cdot Nu_{d_A 0} \cdot \frac{\lambda_l}{d_A} \qquad (6.13)$$

For the reference heat transfer coefficient α_0 of water with Equations (6.9) to (6.13) we receive the value of 6398 W/(m² K). The experimentally determined value is 5600 W/(m² K). For some of the refrigerants the agreement is better.

Table 6.1 contains calculated and experimentally determined reference heat transfer coefficients of water, Freon R134a and propane at $p^* = 0.1$. Further values can be found in VDI Heat Atlas.

Table 6.1: Reference heat transfer coefficient at $p^* = 0,03$

	p_{krit} bar	λ_{l0} W/(m K)	ρ_{l0} kg/m³	c_{pl0} J/(kg K)	$\lambda_{l0} \cdot \rho_{l0} \cdot c_{pl0}$ kg²/(s⁵ K²)	α_0 W/(m² K)	α_{0exp} W/(m² K)
Water	220.64	0.650	843.5	4594	$2.519 \cdot 10^6$	6 398	5 600
R134a	40.60	0.088	1 263.1	1368	$0.154 \cdot 10^6$	3 635	4 500
Propane	42.40	0.108	533.5	2476	$0.143 \cdot 10^6$	3 975	4 000

The excess temperature at which nucleate boiling starts is the temperature at which the heat transfer coefficient of nucleate boiling is larger than that of free convection.

Figure 6.3 shows the transfer from free convection to nucleate boiling of water at 6.62 bar on horizontal tubes of 15 mm outer diameter. In this example the transfer to nucleate boiling is at an excess temperature of 1.5 K.

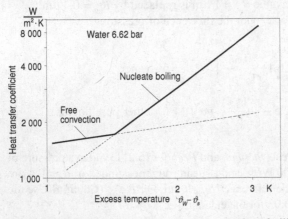

Figure 6.3: Transfer from free convection to nucleate boiling

EXAMPLE 6.1: Water boiling in a pot

Water is brought to nucleate boiling on a heating surface with a power of 2.5 kW in a pot with 250 mm diameter. The pressure is 0.98 bar. The mean roughness index is 0.5 μm.

Material properties of water at $p = 0.03 \cdot p_{krit} = 6.6192$ bar:
$\rho_{l0} = 904.8$ kg/m³, $\rho_{g0} = 3.477$ kg/m³, $c_{p0} = 4.346$ kJ/(kg K), $\sigma_0 = 0.046$ N/m, $Pr_{l0} = 1.07$, $\lambda_{l0} = 0.679$ W/(m K), $\nu_{l0} = 0.185 \cdot 10^{-6}$ m²/s, $T_s = 435.85$ K, $\beta° = 45°$, $r = 2073$ kJ/kg, $a_l = 1.727 \cdot 10^{-7}$ m²/s.

Material properties of water at $p^* = 0.1$:
$\rho_l = 843.5$ kg/m³, $\lambda_l = 0.650$ W/(m K), $c_p = 4.594$ kJ/(kg K)

Material properties of water at $p^* = 0.004444$:
$\rho_l = 958.6$ kg/m³, $\lambda_l = 0.679$ W/(m K), $c_p = 4.216$ kJ/(kg K).

Find

The heat transfer coefficient and excess temperature.

Solution

Assumption

• The temperature of the heating surface is constant.

Analysis

The reference value of the heat transfer coefficient can be taken from Table 6.1. To check it we calculate the reference heat transfer coefficient.

At $p^* = 0.03$ the bubble uncoupling diameter d_{A0} according to Equation (6.6):

$$d_{A0} = 0.0149 \cdot \beta° \cdot \sqrt{\frac{2 \cdot \sigma}{g \cdot (\rho_l - \rho_g)}} =$$

$$= 0.0149 \cdot 45 \cdot \sqrt{\frac{2 \cdot 0.046 \cdot \text{N/m}}{9.806 \cdot \text{m/s}^2 \cdot (904.8 - 3.477) \cdot \text{kg/m}^3}} = 2.163 \text{ mm}$$

The *Nußelt* number Nu_{dA0}, requested to determine α_0 is calculated with Equation (6.12).

$$Nu_{d_A 0} = 0.1 \cdot \left(\frac{\dot{q}_0 \cdot d_{A0}}{\lambda_{l0} \cdot T_{s0}} \right)^{0.674} \cdot \left(\frac{\rho_{g0}}{\rho_{l0}} \right)^{0.156} \cdot \left(\frac{r \cdot d_{A0}^2}{a_{l0}^2} \right)^{0.371} \cdot \left(\frac{a_{l0}^2 \cdot \rho_{l0}}{\sigma_0 \cdot d_{A0}} \right)^{0.35} \cdot Pr_{l0}^{-0.16} = 13.85$$

For the calculation the material properties of water at 6.6192 bar were inserted. The heat flux was as defined as 20 000 W/m².

The function for the pressure correction is determined with Equation (6.11).

$$f(0.03) = 1.73 \cdot 0.03^{0,27} + \left(6,1 + \frac{0.68}{1-0.03}\right) \cdot 0.03^2 = 0.677$$

For the reference heat transfer coefficients α_0 we receive:

$$\alpha_0 = \frac{Nu_{d_{A0}} \cdot \lambda_{l0}}{f(0.03) \cdot d_{A0}} = \frac{13.85 \cdot 0.679 \cdot W/(m \cdot K)}{0.677 \cdot 0.002163 \cdot m} = 6398 \frac{W}{m^2 \, K}$$

This value is the same as that listed in Table 6.1.

The heat transfer coefficient α_B we determine with Equation (6.9). First the heat flux must be calculated:

$$\dot{q} = \frac{\dot{Q}}{A} = \frac{4 \cdot \dot{Q}}{\pi \cdot d^2} = \frac{4 \cdot 2\,500 \cdot W}{\pi \cdot 025^2 \cdot m^2} = 50\,930 \, \frac{W}{m^2}$$

For the material property function we receive:
$$\lambda_{l0} \cdot \rho_{l0} \cdot c_{pl0} = 0.650 \cdot 843.5 \cdot 4594 = 2.519 \cdot 10^6$$
$$\lambda_l \cdot \rho_l \cdot c_{pl} = 0.679 \cdot 958.6 \cdot 4216 = 2.744 \cdot 10^6$$

The pressure correction function delivers:

$$f(0.00444) = 1.73 \cdot 0.00444^{0,27} + \left(6.1 + \frac{0.68}{1-0.00444}\right) \cdot 0.00444^2 = 0.401$$

The heat transfer coefficient with Equation (6.12) is:

$$\alpha_B = \alpha_0 \cdot f(p^*) \cdot \left(\frac{\lambda_l \cdot \rho_l \cdot c_{pl}}{\lambda_{l0} \cdot \rho_{l0} \cdot c_{pl0}}\right)^{0.25} \cdot \left(\frac{R_a}{R_{a0}}\right)^{0.133} \cdot \left(\frac{\dot{q}}{\dot{q}_0}\right)^{0.9-0.3 \cdot p^*} = 6\,271 \frac{W}{m^2 \cdot K}$$

The excess temperature can be determined with the heat flux:

$$\Delta \vartheta = \vartheta_w - \vartheta_s = \frac{\dot{q}}{\alpha_B} = \frac{50\,930 \cdot W/m^2}{6\,271 \cdot W/(m^2 \cdot K)} = 8.12 \, K$$

Discussion

To transfer the given flux an excess temperature of 8.12 K is established. Using the slightly smaller experimental reference values, the value of the heat transfer coefficient would be 14 % smaller, the required excess temperature would be 10 K.

EXAMPLE 6.2: Determine the performance of an electrically heated boiler

With an electric boiler of 6 kW power, steam should be generated at 2 bar pressure. The heating rod has a steel shell of 12 mm diameter and 1 m length. The mean roughness index is 1.5 μm.

The reference values at $p* = 0.03$ can be taken from Example 6.1.

Material properties at $p* = 0.00906$: $r = 2201.6$ kJ/kg, $\rho_l = 942.9$ kg/m³, $\lambda_l = 0.683$ W/(m K), $c_p = 4.247$ kJ/(kg K).

Find

The steam mass flow rate, heat transfer coefficient and excess temperature.

Solution

Assumptions

- The wall temperature is constant.
- The generator is fed with saturated water.

Analysis

The mass flow rate of the steam can be determined by Equation (6.1).

$$\dot{m} = \frac{\dot{Q}}{r} = \frac{6 \cdot \text{kW}}{2201.6 \cdot \text{kJ/kg}} = 0.00273 \text{ kg/s} = \mathbf{9.81\,kg/h}$$

The reference value of the heat transfer coefficient is that of Example 6.1.

Material properties: $\lambda_l \cdot \rho_l \cdot c_{pl} = 0.683 \cdot 942.9 \cdot 4247 = 2.735 \cdot 10^6$

Pressure correction function: $f(0.00906) = 0.486$

For the heat flux, heat transfer coefficient and excess temperature we receive:

$$\dot{q} = \frac{\dot{Q}}{\pi \cdot d \cdot l} = \frac{6 \cdot \text{kW}}{\pi \cdot 0.012 \cdot \text{m} \cdot 1 \cdot \text{m}} = 159\,155 \text{ W/m}^2$$

$$\alpha_B = f(p*) \cdot \left(\frac{\lambda_l \cdot \rho_l \cdot c_{pl}}{\lambda_{l0} \cdot \rho_{l0} \cdot c_{pl0}} \right)^{0.25} \cdot \left(\frac{R_a}{R_{a0}} \right)^{0.133} \cdot \left(\frac{\dot{q}}{\dot{q}_0} \right)^{0.9-0.3 \cdot p*} \cdot \alpha_0 =$$

$$= 0.486 \cdot \left(\frac{2.735}{2.670} \right)^{0.25} \cdot \left(\frac{1.5}{0.4} \right)^{0.133} \cdot \left(\frac{159155}{20000} \right)^{0.9027} \cdot 4346 \ \frac{\text{W}}{\text{m}^2 \cdot \text{K}} = \mathbf{16\,306} \ \frac{\text{W}}{\text{m}^2 \cdot \text{K}}$$

$$\Delta\vartheta = \vartheta_W - \vartheta_s = \frac{\dot{q}}{\alpha_B} = \frac{159155 \cdot \text{W/m}^2}{16306 \cdot \text{W/(m}^2 \cdot \text{K)}} = \textbf{9.76 K}$$

Discussion

The high heat flux generates an intensive bubble generation, providing a very high heat transfer coefficient. Therefore the required excess temperature remains rather low.

EXAMPLE 6.3: Design of an electrically heated boiler

For the start up of a steam power station the feed water storage tank must be heated to increase the pressure to 10 bar. After reaching this pressure, further steam is required to deliver 1.5 kg/s steam for the auxiliary steam lines. The heating is provided with six electric heater rods of 100 mm diameter. The mean roughness index is 3 μm. The excess temperature is 6 K. The reference values at $p* = 0.03$ can be used from Example 6.1.

Material properties at $p = 10$ bar:
$r = 2014.4$ kJ/kg, $\rho_l = 887.1$ kg/m^3, $\lambda_l = 0.673$ W/(m K), $c_p = 4.405$ kJ/(kg K).

Find

Determine the required heat rate and the required heated length of the heaters.

Solution

Assumptions

- The wall temperature is constant.
- Water flowing to the heater rods is at saturation temperature.

Analysis

The heat rate calculated with Equation (6.1) is:

$$\dot{Q} = \dot{m} \cdot r = 1.5 \cdot \frac{\text{kg}}{\text{s}} \cdot 2014.4 \cdot \frac{\text{kJ}}{\text{kg}} = \textbf{3.022 MW}$$

Per heater rod heat rate of 504 kW is required. The material properties are:
$\lambda_l \cdot \rho_l \cdot c_{pl} = 0.673 \cdot 887.1 \cdot 4405 = 2.669 \cdot 10^6 \cdot \text{kg}^2/(\text{s}^5 \text{ K}^2)$.

The relative pressure $p*$ is 0.0452. The correction function for the pressure was determined as:

$$f(0.0453) = 1.73 \cdot 0.0453^{0.27} + \left(6.1 + \frac{0.68}{1-0.0453}\right) \cdot 0.0453^2 = 0.764$$

As the heater rod length is not known, the heat flux will be replaced in Equation (6.9) by the heat transfer coefficient. First we rearrange Equation (6.9).

$$\frac{\alpha_B}{\alpha_0} = f(p*) \cdot \left(\frac{\lambda_l \cdot \rho_l \cdot c_{pl}}{\lambda_{l0} \cdot \rho_{l0} \cdot c_{pl0}}\right)^{0.25} \cdot \left(\frac{R_a}{R_{a0}}\right)^{0.133} \cdot \left(\frac{\dot{q}}{\dot{q}_0}\right)^{0.9-0.3 \cdot p*} = 0.9991 \cdot \left(\frac{\dot{q}}{\dot{q}_0}\right)^{0.8864}$$

The heat flux can be replaced by $\alpha_B \cdot \Delta\vartheta$. This results in the following relationships:

$$\left(\frac{\alpha_B}{\alpha_0}\right)^{0.1-0.3 \cdot p*} = 0.9991 \cdot \left(\frac{\Delta\vartheta}{\Delta\vartheta_0}\right)^{0.8846} \qquad \frac{\alpha_B}{\alpha_0} = 0.9992 \cdot \left(\frac{\Delta\vartheta}{\Delta\vartheta_0}\right)^{7.8028}$$

With the reference values of the heat flux and heat transfer coefficient the reference excess temperature can be determined.

$$\Delta\vartheta_0 = \dot{q}_0 / \alpha_0 = (20\,000 / 4\,346) \cdot \text{K} = 4.602 \text{ K}$$

Heat transfer coefficient, required heating surface area and heated length of the rods can be calculated.

$$\alpha_B = (\Delta\vartheta / \Delta\vartheta_0)^{7.8028} \cdot \alpha_0 = (6 / 3{,}12)^{7.8028} \cdot 4\,346 \cdot \text{W/(m}^2 \cdot \text{K)} = 34\,435 \text{ W/(m}^2 \cdot \text{K)}$$

$$A = \dot{Q} / \dot{q} = \dot{Q} / (\alpha_B \cdot \Delta\vartheta) = 504 \cdot \text{kW} / (34\,435 \cdot 6 \cdot \text{kW/m}^2) = \mathbf{2.436\,m^2}$$

$$l = A / (\pi \cdot d) = 2.436 \cdot \text{m}^2 / (\pi \cdot 0.1 \cdot \text{m}) = \mathbf{7.754\,m}$$

Discussion

The very high heat flux increases the heat transfer coefficient. The heat transfer requires an excess temperature of only 6 K. At higher surface area the heat flux would be reduced and so the heat transfer coefficient, but the excess temperature would increase. The heat transfer coefficients are increasing at this pressure with almost the eighth power of the excess temperature. Just a small decrease of the excess temperature leads to a large increase of the required surface area. In this exam-ple the reduction of the excess temperature from 6 K to 5 K would require a five times larger surface area with 12.13 m². Typical for nucleate boiling is a very rapid increase of the heat transfer coefficients with increasing excess temperature.

6.2 Boiling at forced convection

Boiling can occur in liquids flowing in channels or at impinged surfaces, e.g. at tube bundles. The fluid entering the boiler can be sub-cooled or saturated liquid or a liquid/steam two-phase mixture. At the outlet of the boiler the steam can be wet, saturated or superheated steam. The heat transfer process in a boiler can proceed in a single phase sub cooled or super-heated fluid or in a two phase steam/liquid flow [6.3].

In a flowing fluid the heat transfer first works as described in Chapter 3. Already in the sub-cooled fluid steam bubbles can be generated, which then recondense. However, the steam bubbles influence the heat transfer coefficients in the sub-cooled liquid. With increasing liquid temperature nucleate boiling starts and depending on the velocities a two-phase flow convective heat transfer process develops.

6.2.1 Sub-cooled boiling

The following asymptotic correlation was found for the *sub-cooled boiling*:

$$\alpha = \sqrt[1.2]{\alpha_k^{1.2} + \alpha_B^{1.2}} \qquad (6.14)$$

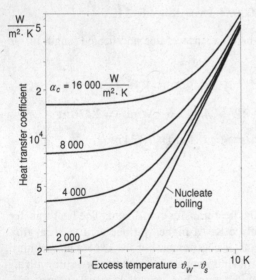

Figure 6.4: Heat transfer coefficients at sub-cooled boiling

Therein α_k is the heat transfer coefficient of forced convection as given in Chapter 3 and α_B that at nucleate boiling according to Equation (6.8).

Figure 6.4 shows an example for heat transfer coefficients at the interference of sub-cooled boiling and forced convection.

6.2.2 Convection boiling

If the heat transfer coefficient of forced convection is larger than that of the nucleate boiling *convection boiling* occurs. It is also called *silent boiling*. With increasing steam velocity the heat transfer coefficient of the convection boundary layer is so large that the excess temperature cannot activate nucleate sites. The evaporation occurs on the liquid surface. Different correlations were found for horizontal and vertical tubes and channels [6.1, 6.2]. They can be applied for all steam qualities between $x = 0$ to $x = 1$.

The ratio of boiling to the liquid heat transfer coefficient in vertical tubes and channels is defined as:

$$\varphi(x) = \frac{\alpha_x}{\alpha_{lo}} = \left\{ \begin{array}{l} (1-x)^{0.01} \cdot \left[(1-x)^{1.5} + 1,9 \cdot x^{0.6} \cdot R^{0.35} \right]^{-2.2} + \\[2mm] + x^{0.01} \cdot \left[\frac{\alpha_{g0}}{\alpha_{l0}} \left(1 + 8 \cdot (1-x)^{0.7} \cdot R^{0.67} \right) \right]^{-2} \end{array} \right\}^{0.5} \qquad (6.15)$$

Therein $R = \rho_l / \rho_g$ is the ratio of liquid to the vapor density. The heat transfer coefficients of the liquid and vapor phase are α_{lo} and α_{g0}. They are calculated as if each phase would flow alone in the tube or channel. The correlations as given in Chapter 3 come to application. The *Reynolds* numbers are:

$$Re_l = \frac{c_{0l} \cdot d_h}{v_l} = \frac{\dot{m} \cdot d_h}{A \cdot \eta_l} \qquad Re_g = \frac{c_{0g} \cdot d_h}{v_l} = \frac{\dot{m} \cdot d_h}{A \cdot \eta_g} \qquad (6.16)$$

Therein is A the cross-section of channel and d_h its hydraulic diameter.
The correlation for horizontal tubes is similar:

$$\varphi(x) = \frac{\alpha_x}{\alpha_{l0}} = \left\{ \begin{array}{l} (1-x)^{0.01} \cdot \left[(1-x)^{1.5} + 1.2 \cdot x^{0.4} \cdot R^{0.37} \right]^{-2.2} + \\[2mm] + x^{0.01} \cdot \left[\frac{\alpha_{g0}}{\alpha_{l0}} \left(1 + 8 \cdot (1-x)^{0.7} \cdot R^{0.67} \right) \right]^{-2} \end{array} \right\}^{0,5} \qquad (6.17)$$

Equation (6.17) takes into account the distribution of the phases in a horizontal tube.

The given correlations are based on tests in circular and rectangular tubes as well as in annular gaps. Figure 6.5 shows the behavior of the heat transfer coefficients in a vertical tube at different density ratios versus the steam quality.

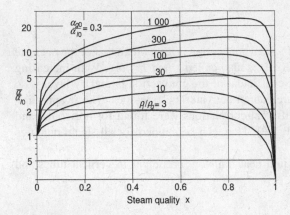

Figure 6.5: Local heat transfer coefficients in vertical tubes vs. vapor quality

Mean heat transfer coefficients can be found by integration of Equations (6.15) and (6.17).

$$\overline{\alpha} = \frac{1}{x_2 - x_1} \cdot \int_{x_1}^{x_2} \alpha(x)dx \qquad (6.18)$$

Figure 6.6 shows the mean heat transfer coefficients at total evaporation from $x = 0$ to $x = 1$ in vertical and Figure 6.7 in horizontal tubes versus of density ratio for different ratios of the liquid to vapor heat transfer coefficient.

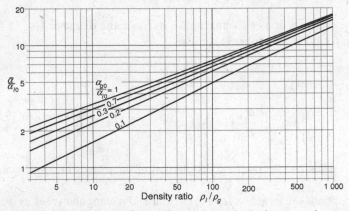

Figure 6.6: Mean heat transfer coefficients in vertical tubes at total evaporation

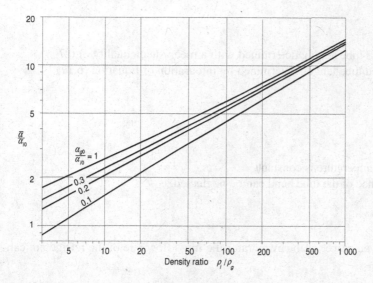

Figure 6.7: Mean heat transfer coefficients in horizontal tubes at total evaporation

In boilers of heat pumps and refrigerators a two-phase mixture enters the evaporator and the liquid part is than completely evaporated. In the boilers of steam generators in steam power plants and in the evaporators of desalinating plants liquid enters the boiler and is not completely evaporated. This is required to avoid the deposit of salt on the boiler surfaces. In these cases Equations (6.15) and (6.17) must be integrated or calculated step-by-step with local values of the mean steam quality. The integration of the equations cannot be performed analytically, but programs like *Mathcad* or *Maple* can deliver the results.

Values determined with a mean steam quality in most cases deliver results with sufficient accuracy.

EXAMPLE 6.4: Design of an evaporator for a refrigerator

The evaporator tubes of a refrigerator have 6 mm internal diameter. The refrigerant R134a evaporates at a pressure of 2 bar. The outside heat resistance related to the internal diameter is $0.9 \cdot 10^{-3}$ (m^2 K)/W. The refrigerant enters the evaporator with a steam quality of 0.4 and is completely evaporated. The heat rate shall be 700 W. The tubes are positioned horizontally. The evaporation temperature is -10.07 °C, the temperature outside the tubes 4 °C.

The latent heat of evaporation of R134a is: $r = 205.88$ kJ/kg.

Condensate properties: $\rho_l = 1327.7$ kg/m^3. $\lambda_l = 0.0971$ W/(m K), $Pr_l = 4.23$, $\eta_l = 0.3143 \cdot 10^{-3}$ kg/(m s).

Vapor properties: $\rho_g = 10.02$ kg/m^3, $\lambda_g = 0.0111$ W/(m K), $Pr_g = 0.609$, $\eta_g = 0.0112 \cdot 10^{-3}$ kg/(m s).

Find

a) The required tube length, determined with a mean steam quality of 0.7.
b) The required tube length, determined by integration of Equation (6.17).

Solution

Assumptions

- The wall temperature is constant.
- The influence of the tube bend can be neglected.

Analysis

With the given power of the evaporator the mass flow rate of the refrigerant can be determined.

$$\dot{m} = \frac{\dot{Q}}{h'' - x_1 \cdot r - h'} = \frac{\dot{Q}}{(x_1 - 1) \cdot r} = \frac{0.700 \text{ kW}}{0.6 \cdot 205.88 \text{ kJ/kg}} = 5.667 \cdot 10^{-3} \frac{\text{kg}}{\text{s}}$$

First the heat transfer coefficients of the liquid and vapor flow will be calculated.

$$Re_l = \frac{4 \cdot \dot{m} \cdot d}{\pi \cdot d_i^2 \cdot \eta_l} = 3\,826 \qquad\qquad Re_g = \frac{4 \cdot \dot{m} \cdot d}{\pi \cdot d_i^2 \cdot \eta_g} = 107\,368$$

$$\xi_l = [1.8 \cdot \log(Re_l) - 1.5]^{-2} = 0.0408 \qquad \xi_g = [1.8 \cdot \log(Re_g) - 1.5]^{-2} = 0.0175$$

$$Nu_l = \frac{\xi_l / 8 \cdot Re_l \cdot Pr_l}{1 + 12.7 \cdot \sqrt{\xi_l / 8} \cdot (Pr_l^{2/3} - 1)} = 33.50$$

$$Nu_g = \frac{\xi_g / 8 \cdot Re_g \cdot Pr_g}{1 + 12.7 \cdot \sqrt{\xi_g / 8} \cdot (Pr_g^{2/3} - 1)} = 171.94$$

$$\alpha_l = \frac{Nu_l \cdot \lambda_l}{d_i} = 542.1 \frac{\text{W}}{\text{m}^2 \text{ K}} \qquad\qquad \alpha_g = \frac{Nu_g \cdot \lambda_g}{d_i} = 318.1 \frac{\text{W}}{\text{m}^2 \text{ K}}$$

$$\alpha_g / \alpha_l = 0.587$$

a) First the ratio of the two-phase flow to the liquid heat transfer coefficient will be calculated with Equation (6.17) for a steam quality of $x = 0.7$.

$$\varphi(x) = \frac{\alpha_x}{\alpha_l} = \left\{ \begin{array}{l} (1-x)^{0.01} \cdot \left[(1-x)^{1.5} + 1.2 \cdot x^{0.4} \cdot R^{0.37} \right]^{-2.2} + \\ + x^{0.01} \cdot \left[\dfrac{\alpha_g}{\alpha_l} \left(1 + 8 \cdot (1-x)^{0.7} \cdot R^{0.67} \right) \right]^{-2} \end{array} \right\}^{-0.5} = 2.206$$

The mean heat transfer coefficient:

$$\bar{\alpha} = \alpha_l \cdot \varphi(0.7) = 2.206 \cdot 542.1 \frac{W}{m^2\ K} = 1196.1 \frac{W}{m^2\ K}$$

Overall mean heat transfer coefficient related to the inner diameter:

$$\bar{k} = \left(\frac{1}{\bar{\alpha}} + R_a \right)^{-1} = \left(\frac{1}{1196.1} + 0.9 \cdot 10^{-3} \right)^{-1} \cdot \frac{W}{m^2\ K} = 576.0 \frac{W}{m^2\ K}$$

The surface required for the transfer of the heat rate is:

$$A = \frac{\dot{Q}}{k \cdot (\vartheta_a - \vartheta_i)} = \frac{700 \cdot W \cdot m^2 \cdot K}{576.0 \cdot W \cdot (4 + 10.07) \cdot K} = 0.086\ m^2$$

For the required tube length we receive:

$$l = A / (\pi \cdot d_i) = \textbf{4.582 m}$$

b) The integral of Equation. (6.16) can be calculated e.g. with *MathCad*. We receive:

$$\bar{\varphi}(x) = \frac{1}{x_2 - x_1} \cdot \int_{x_1}^{x_2} \varphi(x) \cdot dx = \frac{1}{1 - 0.4} \cdot \int_{0.4}^{1} \varphi(x) \cdot dx = 2.113$$

The integrated mean heat transfer coefficient has a 4.4 % lower value as calculated with the mean steam quality and would require a 2.1 % longer tube.

Discussion

In this example the heat transfer coefficient is 2.2 times larger compared to the liquid flow convective heat transfer. The case discussed here delivers a rather good agreement for both calculations, i.e. calculation with the mean steam quality delivers almost the same heat transfer coefficient as received with the exact integration. Step-by-step calculations with three to five steps results usually very close to the integrated value.

7   Thermal radiation

Heat transfer by *thermal radiation* is carried by electromagnetic waves. Contrary to heat conduction, where the transfer is managed by the movement of molecules, atoms or electrons, i.e. a transfer medium is required, radiation does not need a medium and thus can even occur in vacuum. Heat transfer by radiation happens in vacuum, or in materials (glasses and gases) which allow the transmission of electromagnetic waves. In the second case the radiation goes along with conduction and both heat rates have to be calculated separately and added. Furthermore, gases consisting of molecules with more than two atoms may emit or absorb electromagnetic waves.

> *Radiation occurs from the surface of solid or liquid bodies and also from gases with molecules having more than two atoms.*

The wavelength of electromagnetic waves transferring heat varies between 0.8 and 400 μm. Electromagnetic waves in this wavelength region are called ultra red light. The wavelength range of visible light is between 0.35 and 0.75 μm. At low temperatures the part of visible radiation is so low that it cannot be detected by human eyes. At high temperatures the part of visible light is increasing and can be detect by our eyes (e.g. the glowing filament of an electric bulb).

The intensity of thermal radiation increases with increasing temperature but also at very low temperatures the thermal radiation can be of importance as e.g. at the insulation of cryogenic systems.

Depending on its surface characteristic, radiation waves will be fully or partially reflected, transmitted or absorbed at the surface of a body. The ratio of absorbed to total radiation is α, the ratio of transmitted vs. total is τ and the ratio of reflected vs. total radiation is ρ. α is also called *absorptivity*. The sum of the three ratios is always 1.

$$\alpha + \rho + \tau = 1 \tag{7.1}$$

Solids and liquids block the transmission of electromagnetic waves even at very small thicknesses: metals at 1 μm and liquids at 1 mm. For almost all bodies the part of reflection, absorption and transmission are dependent on the wavelength.

> *Every body, with a temperature higher than absolute zero, emits electromagnetic waves.*

The capability of emission is a characteristic of the body. A so called *black body* is capable of emitting electromagnetic waves at a certain temperature with maximum intensity. The capability of other bodies to emit electromagnetic waves at the same temperature is determined by the *emissivity* ε. The emissivity is the ratio of the emission intensity of the body at a certain temperature to that of the black body at the same temperature.

The *Kirchhoff's law* states:

> *The emissivity ε of a body at stationary conditions has the same value as its absorptivity α.*

$$\varepsilon = \alpha \tag{7.2}$$

Depending on their *reflectivity*, *absorptivity* and *transmissivity* bodies are assigned with the following characteristics respectively names:

black: the radiation will be completely absorbed ($\alpha = \varepsilon = 1$)
white: the radiation will be completely reflected ($\rho = 1$)
gray: the absorptivity for all wavelengths is the same ($\varepsilon < 1$)
colored: certain wavelengths (those of the colors) are preferentially reflected
reflective: all rays are reflected with the same angle as the inlet angle
soft/diffuse: the radiation is reflected diffuse in all directions.

7.1   Basic law of thermal radiation

A black body can be approximated by a hollow volume with adiabatic, isothermal inner walls and a small opening through which radiation enters and leaves. The radiation emitted solely through the opening, is called *black radiation*.

The *spectral specific intensity* of the black radiation $i_{\lambda,s}$ is given by *Planck's radiation law*.

$$i_{\lambda,s} = \frac{C_1}{\lambda^5 \cdot (e^{C_2/(\lambda \cdot T)} - 1)} \tag{7.3}$$

The consonants C_1 and C_2 are given as:

$$C_1 = 2 \cdot \pi \cdot c^2 \cdot h = 3.7418 \cdot 10^{-16} \ \text{W} \cdot \text{m}^2$$
$$C_2 = c \cdot h / k = 1.438 \cdot 10^{-2} \ \text{K} \cdot \text{m} \tag{7.4}$$

Both constants do not include empiric terms but only physical constants. They are the light velocity c, the *Planck constant h*, the *Boltzmann constant k*. The values of these physical constants are:

$c = 299\,792\,458$ m/s, $h = 6.6260755 \cdot 10^{-34}$ J $\cdot$ s, $k = 1.380641 \cdot 10^{-23}$ J/K.

The term $i_{\lambda,s}$ is the radiation intensity (flux) of a black radiator (the index s is for black radiation) divided by the wavelength, at which the radiation occurs. As the intensity is given per wavelength, its unit is W/m^3.

Figure 7.1 shows the spectral specific intensity at different absolute temperatures vs. wavelength. As the diagram shows, the radiation has a maximum for each temperature at a certain wavelength. Differentiation of Equation (7.3) to the wavelength and set to zero, delivers the wavelength of the maximum as a function of the temperature.

$$\lambda_{i=max} = 2\,898 \ \mu m \cdot K/T \tag{7.5}$$

With increasing temperature the maximum is displaced to smaller wavelengths. This relationship is called *Wien's displacement law*.

The temperature of the sun's surface is approximately 6 000 K. The maximum is there at a wavelength of 0.48 μm, i.e. in the visible range.

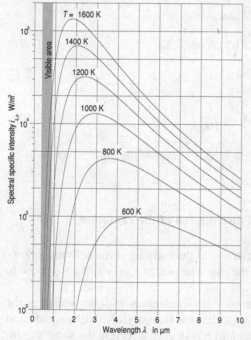

Figure 7.1: Spectral specific intensity of black radiation vs. wavelength

7.2 Determination of the heat flux of radiation

The intensity of radiation, which is the heat flux, can be determined by integration of Equation (7.3) from zero to infinite wavelength.

$$\dot{q}_s = \int\limits_{\lambda=0}^{\lambda=\infty} i_{\lambda,s} \cdot d\lambda = \frac{2 \cdot \pi^5 \cdot k^4}{15 \cdot h^4 \cdot c^2} \cdot T^4 = \sigma \cdot T^4 \qquad (7.6)$$

The *Stefan-Boltzmann-constant* σ, can be calculated with the physical constants given before:

$$\sigma = (5.6696 \pm 0.0075) \cdot 10^{-8} \ \text{W} \cdot \text{m}^{-2} \cdot \text{K}^{-4} \qquad (7.7)$$

This is presently the most exact value, based on measurements of the physical constants. In practice the value of $5.67 \cdot 10^{-8} \ \text{W} \cdot \text{m}^{-2} \cdot \text{K}^{-4}$ is commonly used. For a simpler calculation, the following equation is proposed:

$$\dot{q}_s = C_s \cdot \left(\frac{T}{100}\right)^4 \qquad (7.8)$$

Here C_s is the *radiation constant of black bodies*.

$$C_s = 10^8 \cdot \sigma = 5.67 \ \text{W} \cdot \text{m}^{-2} \cdot \text{K}^{-4} \qquad (7.9)$$

For non-black bodies the heat flux is:

$$\dot{q} = \varepsilon \cdot C_s \cdot \left(\frac{T}{100}\right)^4 \qquad (7.10)$$

The heat flux emitted from a non-black body is that of the black body multiplied by the emissivity.

7.2.1   Intensity and directional distribution of the radiation

The emission changes with the characteristic of the emitting surface. In the following gray bodies will be discussed, as the characteristic of a gray surface is close to that of most technical surfaces.

The intensity of a dot shaped radiation source decreases with the square of the distance. The *Lambert's cosine law* says, that the intensity of a diffuse radiation emitted from an infinite surface element dA has in each direction the same magnitude. However the emittance decreases proportionally to the cosinus of the angle to the orthogonal to the surface.

$$\dot{q}_\beta = \dot{q}_n \cdot \cos \beta \qquad (7.11)$$

The integration over a half sphere delivers the total emission to the space:

$$\dot{q} = \dot{q}_n \cdot \pi \qquad (7.12)$$

From a gray surface the heat flux radiated to a half sphere is π times the heat flux of the orthogonal ration. For non-gray bodies this law is only approximately correct. The emissivity of metal surfaces increases with radiation angel and for non-metallic surfaces it decreases.

7.2.2 Emissivities of technical surfaces

The emissivity of a surface depends on the surface characteristic and the temperature. Ageing ,for example, due to contamination, oxidation and corrosion can lead to strong alteration of the emissivity. The exact value of the emissivity of a technical surface can only be determined by measurements. Conclusions, regarding the emissivities from optical or other evaluations may lead to completely wrong results.

Table 7.1: Emission coefficients of technical surfaces (Source: [7.2])

Material	State	Temperature °C	ε_n	ε
Aluminum	rolled shiny	170	0.039	0.049
		900	0.060	
	strongly oxidized	90	0.020	
		504	0.310	
Aluminum oxide		277	0.630	
		830	0.260	
Copper	polished	20	0.030	
	slightly tarnished	20	0.037	
	black oxidized	20	0.780	
Iron, steel	polished	430	0.144	
	cast	100	0.800	
Steel	oxidized	200	0.790	
Tungsten		25		0.024
		1000		0.150
		3000		0.450
Glass		20	0.940	
Gypsum		20	0.850	
Brick		20	0.930	
Wood (Oak)		20		0.900
Paint	black, flat	80		0.920
Paint	white	100		0.940
Radiator paint (VDI-74)		100	0.925	
Water		0	0.950	
		100	0.960	
Ice		0	0.966	

The normal component of the emissivity can simply be measured between two parallel plates. For this reason in literature commonly the normal components ε_n of the emission coefficients are published. With the diagram in Figure 7.2 the emission coefficient of the total radiation can determined.

In practice the emission coefficients, depending on the state of the surface may have large deviations from the values reported in literature.

Further data can be found in VDI Heat Atlas [7.2] and W. Wagner: Wärme-übertragung [7.4], [7.5, 7.6] and in Appendix A11 of this book.

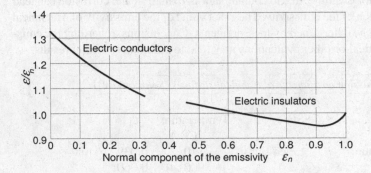

Figure 7.2: Diagram for the determination of the emission coefficient of the total radiation

7.2.3   Heat transfer between two surfaces

In many of the technically interesting cases thermal radiation occurs between two or more surfaces. The transfer phenomena by thermal radiation can easily be explained in the example of interaction between two surfaces. The surface area 1 emits rays according to its temperature T_1 and characteristic of the surface. As defined by the directional distribution a part of the emitted rays hits the second surface and will there be absorbed, reflected or passed through. The second surface in turn emits rays according to its temperature T_2 and characteristics of the surface. These emitted and reflected rays hit the first surface according to their directional distribution.

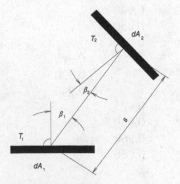

Figure 7.3: Radiation heat transfer between two surfaces

Figure 7.3 shows the thermal radiation between two infinitely small surface elements dA_1 and dA_2 of the surface 1 and 2. The temperature of surface 1 is T_1 and that of the surface 2 T_2. Similarly are the emission coefficients ε_1 and ε_2. *Lambert's law* delivers for the heat flow rate between the two gray diffuse surfaces:

$$\dot{Q}_{12} = \varepsilon_1 \cdot \varepsilon_2 \cdot C_s \cdot \left[\left(\frac{T_1}{100} \right)^4 - \left(\frac{T_2}{100} \right)^4 \right] \cdot \int\limits_{A_1} \int\limits_{A_2} \frac{\cos \beta_1 \cdot \cos \beta_2}{\pi \cdot s^2} \cdot dA_1 \cdot dA_2 \quad (7.13)$$

The double integral in this equation is representing the geometrical characteristics. The integral divided by the surface A_1 delivers the *irradiation coefficient* φ_{12}.

$$\varphi_{12} = \frac{1}{A_1} \cdot \int\limits_{A_1} \int\limits_{A_2} \frac{\cos \beta_1 \cdot \cos \beta_2}{\pi \cdot s^2} \cdot dA_1 \cdot dA_2 \qquad (7.14)$$

As radiation heat transfers in reverse directions must be independent of geometry, the following expression applies:

$$A_1 \cdot \varphi_{12} = A_2 \cdot \varphi_{21} \qquad (7.15)$$

With this the heat rate can be given as:

$$\dot{Q}_{12} = \varphi_{12} \cdot \varepsilon_1 \cdot \varepsilon_2 \cdot C_s \cdot A_1 \cdot \left[\left(\frac{T_1}{100} \right)^4 - \left(\frac{T_2}{100} \right)^4 \right] \qquad (7.16)$$

For pairs of surfaces i and k in thermal radiation interaction the following reciprocal correlation is valid:

$$A_i \cdot \varphi_{ik} = A_k \cdot \varphi_{ki} \qquad (7.17)$$

In case of the thermal radiation interaction of the surface i with other surfaces in the space surrounding i, the energy balance delivers the following relation:

$$\sum_{k=1}^{n} \varphi_{ik} = 1 \qquad (7.18)$$

The problem of determining the heat rate is the evaluation of the integrals in Equation (7.14). For non-gray technical surfaces the determination is even more complicated, as the radiation is partially absorbed and transmitted.

For simple geometrical surfaces of lateral dimensions much larger than its distance the irradiation coefficient can be calculated as:

$$\dot{Q}_{12} = C_{12} \cdot A \cdot \left[\left(\frac{T_1}{100} \right)^4 - \left(\frac{T_2}{100} \right)^4 \right] \tag{7.19}$$

Therein C_{12} is the *thermal radiation exchange coefficient*.

7.2.3.1 *Parallel gray plates with identical surface area size*

Between two parallel gray plates of the same size with the temperatures T_1 and T_2 (Figure 7.4) the thermal radiation exchange coefficient is given as:

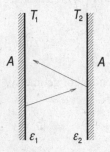

Figure 7.4: Thermal radiation between two parallel gray palates of identical size

$$C_{12} = \frac{C_s}{1/\varepsilon_1 + 1/\varepsilon_2 - 1} \tag{7.20}$$

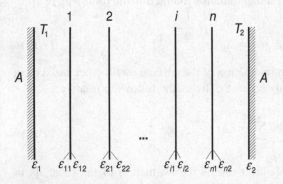

Figure 7.5: Thermal radiation between several parallel gray palates of identical size

For *n* additional parallel plates in between two outer plane plates of identical size as shown in Figure 7.5, an analysis delivers the following correlation:

$$C_{12} = \frac{C_s}{1/\varepsilon_1 + 1/\varepsilon_2 - 1 + \sum_{i=1}^{n}(1/\varepsilon_{i1} + 1/\varepsilon_{i2} - 1)} \tag{7.21}$$

Therein ε_{i1} is the emission coefficient of the i-th plate on the side facing T_1 and ε_{i2} facing T_2. If all n plates have the same emissivity, Equation (7.21) simplifies:

$$C_{12} = \frac{C_s}{1/\varepsilon_1 + 1/\varepsilon_2 - 1 + n \cdot (2/\varepsilon_i - 1)} \tag{7.22}$$

If all surfaces have the same emissivity, a further simplification results:

$$C_{12} = \frac{C_s}{(n+1) \cdot (2/\varepsilon - 1)} \tag{7.23}$$

As the size of all plates is identical, in Equation (7.19) the surface A can be inserted.

Due to the additional plates the value of the denominator in the above equations increases and subsequently the heat flow rate decreases. The additional plates result in improved heat insulation. For the insulation of very cold fluids such as liquid helium, this type of insulation comes to application. The tank to be insulated is surrounded with an outer shell. The space in between is evacuated and provided with polished thin aluminum foils (super insulation).

7.2.3.2 Surrounded bodies

For surrounded bodies like a sphere in a hollow sphere or a cylinder in hollow cylinder (Figure 7.6) the following correlation applies:

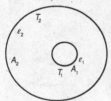

Figure 7.6: The radiation of a surrounded body

$$C_{12} = \frac{C_s}{\dfrac{1}{\varepsilon_1} + \dfrac{A_1}{A_2} \cdot \left(\dfrac{1}{\varepsilon_2} - 1\right)} \tag{7.24}$$

In Equation (7.24) A_1 is always the surface of the surrounded body, i.e. A_1 is smaler than A_2. Here in the case that T_1 is lower than T_2 and the heat flow rate calculated with Equation (7.19) would be negative. This has to be taken in to account.

If the surface A_1 is much smaller than A_2, Equation (7.24) delivers:

$$C_{12} = \varepsilon_1 \cdot C_s \tag{7.25}$$

Even in case of bodies with rather complex geometry surrounded by an essentially larger surface Equation (7.25) delivers results very close to reality (e.g. a radiator).

For a large number of geometries thermal radiation exchange coefficients are given in VDI Heat Atlas [7.1].

EXAMPLE 7.1: Insulation with special emissivity window panes

The window panes of a house are replaced by high reflexivity glass panes. The material properties of the original and high reflexivity glass are:

	original	high reflexivity
Absorptivity α	0.80	0.40
Reflexivity ρ	0.05	0.50
Transmissivity τ	0.15	0.10

The old panes were heated to 35°C on the room side, with the new panes this temperature dropped to 28 °C. The temperature of the walls and the air in the room is 22 °C. The heat transfer coefficient of free convection is 5 W/(m² K). The heat flux of the irradiation of the sun was measured as 700 W/m².

Find

Calculate the heat flux entering the room through the window panes.

Solution

Schematic   See sketch.

Assumptions

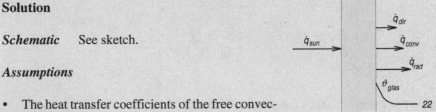

- The heat transfer coefficients of the free convection are constant.
- The glass temperature is constant.

Analysis

The heat flux through the gals of the window panes into the room has three components:

- the sun rays directly passing through the glass,
- the free convection,
- the thermal radiation from the windowpanes glass to the room.

The three heat fluxes will now be calculated for the normal and for the high reflexivity glass.

The portion of heat flux of the sun rays passing directly through is given by the transmissivity.

	original	new

$$\dot{q}_{dir} = \dot{q}_{sun} \cdot \tau \qquad\qquad 105 \text{ W/m}^2 \qquad 70 \text{ W/m}^2$$

The heat flux of the free convection:

$$\dot{q}_{conv} = \alpha_K \cdot (\vartheta_{glass} - \vartheta_R) \qquad\qquad 65 \text{ W/m}^2 \qquad 30 \text{ W/m}^2$$

The heat flux transferred by radiation from the pane to the walls can be determined by Equation (7.25):

$$\dot{q}_{rad} = \varepsilon \cdot C_s \cdot \left[\left(\frac{T_{glass}}{100} \right)^4 - \left(\frac{T_R}{100} \right)^4 \right] \qquad\qquad 65 \text{ W/m}^2 \qquad 14 \text{ W/m}^2$$

For the heat flux from window panes out of normal glass we receive **235 W/m²**, for the high reflexivity glass **114 W/m²**.

Discussion

The high reflexivity glass reflects 50 % of the sun rays hitting the window plane; this is about 45 % more than with the normal glass. The absorptivity is 50 % lower and the window pane will be heated up less, thus the convection heat transfer as well as the thermal radiation to the room will be reduced. The larger part of the heat flux to the room with high reflexivity glass is caused by the convection and radiation from the pane.

EXAMPLE 7.2: Design of a light bulb filament

A 240 Volt light bulb shall have an electric power of 100 W at a filament temperature of 3 100 °C. The specific electric resistance of the tungsten filament has the value of $\rho_{el} = 73 \cdot 10^{-9}$ Ω m. The temperature of the glass of the bulb is 90 °C.

Find

a) The length and diameter of the filament.
b) Efficiency of the electric bulb.

Solution

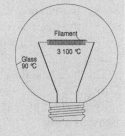

Schematic   See sketch.

Assumptions

- The surface of the glass is much larger than that of the filament and is therefore negligible.
- The influence of the filament holders and the convective heat transfer in the bulb can be neglected.
- The temperature of the filament is constant.

Analysis

a)   The heat flow rate transferred from the filament by thermal radiation is equal to the electric power supplied to the filament. To determine heat flow rate of the thermal radiation the emissivity of tungsten is required. In Table 7.1 it is given as 0.45. The surface area for a heat flow rate of 100 W can be determined by Equation (7.19). The thermal radiation exchange coefficient is with Equation (7.25) $C_{12} = \varepsilon \cdot C_s$. For the surface area required for the 100 W heat flow rate we receive:

$$A = \frac{\dot{Q}}{C_s \cdot \varepsilon \cdot \left\{ \left(\dfrac{T_1}{100} \right)^4 - \left(\dfrac{T_2}{100} \right)^4 \right\}} = 3.028 \cdot 10^{-5} \text{ m}^2$$

For 100 W electric power at a voltage of 240 V the following electric resistance is required:

$$R = \frac{U^2}{P_{el}} = \frac{240^2 \cdot \text{V}^2}{100 \cdot \text{W}} = 576 \ \Omega$$

The surface area A and the electric resistance R are functions of the dimensions of the filament. The following relationships can be given:

$$A = \pi \cdot d \cdot l \qquad\qquad R = \frac{4 \cdot \rho_{el} \cdot l}{\pi \cdot d^2}$$

Now the diameter d of the filament can be determined as:

$$d = \sqrt[3]{\frac{4 \cdot \rho \cdot A}{\pi^2 \cdot R}} = \sqrt[3]{\frac{4 \cdot 73 \cdot 10^{-9} \cdot \Omega \cdot \text{m} \cdot 3.021 \cdot 10^{-5} \cdot \text{m}^2}{\pi^2 \cdot 576 \cdot \Omega}} = \textbf{0.0116 mm}$$

The required length of the filament is:

$$l = \frac{A}{\pi \cdot d} = \frac{3.021 \cdot 10^{-5} \text{m}^2}{\pi \cdot 12 \cdot 10^{-6} \text{m}} = 0.832 \text{ m}$$

b)   The objective of an electric bulb is to provide light, i.e. emit visible light (wavelength 0.35 to 0.75 µm). By integrating Equation (7.3) between these wavelengths and multiplying the result by the emissivity we receive the heat flux of the radiation in the range of the visible light.

$$\dot{q}_{visible} = \int_{\lambda_1}^{\lambda_2} \frac{\varepsilon \cdot C_1}{\lambda^5 \cdot (e^{C_2/(\lambda \cdot T)} - 1)} d\lambda =$$

$$= \int_{\lambda=0.35\,\mu m}^{\lambda=0.75\,\mu m} \frac{0.45 \cdot 3.7418 \cdot 10^{-16} \cdot \text{W} \cdot \text{m}^2}{\lambda^5 \cdot (e^{0.01438 \cdot \text{K} \cdot \text{m} / (\lambda \cdot 3373 \cdot \text{K})} - 1)} d\lambda = 0.5504 \frac{\text{MW}}{\text{m}^2}$$

For the total emitted heat flux by thermal radiation we receive:

$$\dot{q}_{tot} = \frac{P_{el}}{A} = \frac{100 \text{ W}}{2.7185 \cdot 10^{-5} \cdot \text{m}^2} = 3.303 \frac{\text{MW}}{\text{m}^2}$$

The efficiency of the bulb is:

$$\eta_{Bulb} = \dot{q}_{visible} / \dot{q}_{tot} = 0.167$$

Discussion

The temperature of the filament determines the surface area of the filament required to emit a heat rate of 100 W, i.e. the determined surface emits at the given temperature a heat rate of 100 W. The filament is designed such that at a voltage of 240 V its electric resistance provides an electric power of 100 W and the surface area being sufficient for emitting the 100 W heat flow rate by thermal radiation. The relatively long filament can be realized by coils and redirections with multiple holders. The major part of the electric power is emitted as heat to the glass and transferred from there to the environment. The efficiency of 17%, calculated in this example with some simplifying assumptions, cannot be reached in reality. Only about 8% to 12% of the radiated power is in the range of the visible light.

EXAMPLE 7.3: Performance of a radiator

A radiator with a length of 1.2 m, a height of 0.45 m and a thickness of 0.02 m has a surface temperature of 60 °C. Its surface is covered with radiator enamel. The walls of the room have a temperature of 20 °C and the air in the room 22 °C. The material properties of the air: $\lambda = 0.0245$ W/(m K), $v = 14 \cdot 10^{-6}$ m²/s, $Pr = 0.711$.

Find

The heat rate transferred from the radiator by radiation and by free convection.

Solution

Schematic   See sketch.

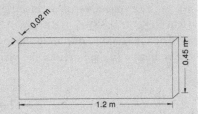

Assumptions

- As the surface area of the room walls is much larger than that of the heater, it can be neglected.
- The thin side walls of the heater are not taken into consideration.
- The temperature of the heater is constant.

Analysis

The heat rate transferred by thermal radiation can be determined with Equations (7.19) and (7.25). The emissivity of the heater surface can be calculated by the value given in Table 7.1 and the diagram shown in Figure 7.2:

$$\varepsilon = \varepsilon_n \cdot 0.96 = 0.925 \cdot 0.96 = 0.89$$

The surface area of the heater: $A = 2 \cdot l \cdot h = 2 \cdot 1.2 \cdot 0.45 \text{ m}^2 = 1.08 \text{ m}^2$

The heat flow rate of thermal radiation is:

$$\dot{Q}_{Str} = \varepsilon \cdot C_s \cdot A \cdot \left[(T_1 / 100)^4 - (T_2 / 100)^4 \right] =$$
$$= 0.89 \cdot 5.67 \cdot \text{W/(m}^2 \cdot \text{K}^4) \cdot 1.08 \cdot \text{m}^2 \cdot (3.332^4 - 2.993^4) \cdot \text{K}^4 = \mathbf{257\ W}$$

The heat rate transferred by free conductivity can be calculated as described in Chapter 4. The *Rayleigh* number is calculated with Equations (4.3) and (4.6).

$$Ra = Gr \cdot Pr = \frac{g \cdot h^3 \cdot (\vartheta_W - \vartheta_0)}{T_0 \cdot v^2} \cdot Pr =$$
$$= \frac{9.806 \cdot \text{m} \cdot \text{s}^{-2} \cdot 0.45^3 \cdot \text{m}^3 \cdot (60 - 22) \cdot \text{K}}{295.15 \cdot \text{K} \cdot 14^2 \cdot 10^{-12} \cdot \text{m}^4 \cdot \text{s}^{-2}} \cdot 0.711 = 417.4 \cdot 10^6$$

The *Nußelt* number is determined with Equations (4.7) and (4.8):

$$f_1(Pr) = \left(1 + 0.671 \cdot Pr^{-9/16}\right)^{-8/27} = \left(1 + 0.671 \cdot 0.711^{-9/16}\right)^{-8/27} = 0.838$$

$$Nu_h = \left\{ 0.852 + 0.387 \cdot Ra^{1/6} \cdot f_1(Pr) \right\}^2 = 94.51$$

For the heat transfer coefficient we receive:

$$\alpha = \frac{Nu_h \cdot \lambda}{h} = \frac{94.51 \cdot 0.0245 \cdot W}{0.45 \cdot m \cdot m \cdot K} = 5.146 \frac{W}{m^2 \cdot K}$$

The heat flow rate transferred by free convection results as:

$$\dot{Q}_{conv} = A \cdot \alpha \cdot (\vartheta_1 - \vartheta_0) = 1.08 \cdot m^2 \cdot 5.146 \cdot W \cdot m^{-2} \cdot K^{-1} \cdot (60 - 22) \cdot K = \mathbf{211\ W}$$

Discussion

In this example the heat rates transferred by thermal radiation and by free convection are compared. The heat flow rate by thermal radiation is larger. The part of the thermal radiation increases stronger with increasing wall temperature than that of the free convection. This is the reason that in earlier days, when the heater surfaces worked at 80 °C, they were called radiators and this name is still used now.

Attention! Here it is important, that for the thermal radiation the ambient temperature is that of the wall and for the free convection the air temperature. The total heat flow rate is the addition of the heat flow rates of thermal radiation and free convection.

EXAMPLE 7.4: Distortion of the temperature measurement by radiation

With a spherical shaped temperature sensor of 2 m diameter the temperature of the exhaust gas of a car is measured. The measured temperatures seem to be too low. As it was assumed that the measurement could be distorted by thermal radiation also the wall temperature of the exhaust pipe was determined. At a measured exhaust gas temperature of 880 °C the wall temperature of the exhaust pipe had a temperature of 250 °C. The emissivity of the sensor is 0.4, the velocity of the gas in the exhaust pipe 25 m/s. The material properties of the exhaust gas:

$\lambda = 0.076$ W/(m K), $v = 162 \cdot 10^{-6}$ m²/s, $Pr = 0.74$.

Find

a) The temperature of the exhaust gas.
b) Show possibilities to improve the accuracy of the measurement.

Solution

Schematic   See sketch.

250 °C

25 m/s →

880 °C

Assumptions

- As the surface area of the exhaust pipe is much larger than that of the sensor, Equation (7.25) applies.
- The measurement is not influenced by the fixation of the sensor.

Analysis

a)   The gas-heated sensor transfers heat to the colder exhaust pipe wall and has therefore a lower temperature than the exhaust gas. As the temperature of the sensor is lower than that of the exhaust gas, heat is transferred from the gas to the sensor by forced convection. The sensor transfers the same amount of heat rate to the exhaust pipe wall.

$$\dot{Q}_{rad} = \dot{Q}_{conv}$$

The heat flow rate transferred by thermal radiation can be determined with Equations (7.19) and (7.25).

$$\dot{Q}_{rad} = \varepsilon \cdot A \cdot C_s \cdot \left[(T_S /100)^4 - (T_W /100)^4 \right]$$

Temperature of the sensor is T_S and of the exhaust pipe T_W. The heat transfer coefficient on a body in the cross-flow is discussed in Chapter 3.2.3. The equation to be applied is defined by the *Reynolds* number.

$$Re_{L'} = c \cdot d / v = 25 \cdot 0.002 / 162 \cdot 10^{-6} = 309$$

Equation (3.30) delivers the *Nußelt* number:

$$Nu_{L',lam} = 0.664 \cdot \sqrt[3]{Pr} \cdot \sqrt{Re_{L'}} = 0.664 \cdot \sqrt[3]{0.74} \cdot \sqrt{309} = 10.55$$

The heat transfer coefficient of the forced convection is:

$$\alpha = \frac{Nu_{L',lam} \cdot \lambda}{d} = \frac{10.55 \cdot 0.076 \cdot W}{0.002 \cdot m \cdot m \cdot K} = 401 \ \frac{W}{m^2 \cdot K}$$

The heat rate transferred to the sensor by forced convection:

$$\dot{Q}_{conv} = \alpha \cdot A \cdot (T_0 - T_S)$$

As both heat rates are equal, the exhaust gas temperature can be determined.

$$T_0 = \frac{\varepsilon \cdot C_s}{\alpha} \cdot \left[(T_S / 100)^4 - (T_W / 100)^4 \right] + T_S =$$

$$= \frac{0.4 \cdot 5.67}{401 \cdot \text{K}^3} \cdot (11.532^4 - 5.232^4) \cdot \text{K}^4 + 1153.2 \text{ K} = \textbf{975.8 °C}$$

The temperature of the exhaust gas is 92.8 K higher than the measured value.

b) The accuracy of the measurement can be improved by insulation of the exhaust pipe wall. With this measure the cooling down of the exhaust pipe by the airstream is avoided. If this is not possible, a shield can be installed around the sensor. This can be e.g. a polished steel pipe of 10 mm diameter, 20 mm length and an emissivity of 0.06. The sensor then has only a radiation heat transfer to the shield which transfers heat by radiation to the exhaust pipe wall. The pipe can be calculated as a good approximation of a plane wall. The characteristic length is the length of the pipe. The *Reynolds* number is:

$$Re_l = c \cdot l / v = 25 \cdot 0.02 / 162 \cdot 10^{-6} = 3086$$

The *Nußelt* number can be determined with Equation (3.21).

$$Nu_{l,lam} = 0.664 \cdot \sqrt[3]{Pr} \cdot \sqrt{Re_l} = 0,664 \cdot \sqrt[3]{0.74} \cdot \sqrt{3086} = 33.42$$

The heat transfer coefficient:

$$\alpha = \frac{Nu_{L',lam} \cdot \lambda}{d} = \frac{33.42 \cdot 0.076 \cdot \text{W}}{0.02 \cdot \text{m} \cdot \text{m} \cdot \text{K}} = 127.0 \ \frac{\text{W}}{\text{m}^2 \cdot \text{K}}$$

The pipe is heated by the exhaust gas from both sides but transfers heat by radiation only from the outer side to the exhaust pipe. For the determination of the heat rate by convection the doubled surface must be used.

$$\dot{Q}_{conv} = \alpha \cdot 2 \cdot A \cdot (T_0 - T_{pipe})$$

For the temperature T_{pipe} we receive:

$$T_0 = \frac{\varepsilon \cdot C_s}{\alpha} \cdot \left[(T_S / 100)^4 - (T_W / 100)^4 \right] + T_S =$$

$$= \frac{0.4 \cdot 5.67}{401 \cdot \text{K}^3} \cdot (11.532^4 - 5.232^4) \cdot \text{K}^4 + 1153.2 \text{ K} = \textbf{975.8 °C}$$

The temperature of the exhaust gas is 27.6 K higher than that of the pipe. The distortion of the sensor temperature can be determined as before and we receive for the temperature difference:

$$T_0 - T_S = \frac{\varepsilon \cdot C_s}{\alpha} \cdot \left[(T_S / 100)^4 - (T_{pipe} / 100)^4 \right] =$$

$$= \frac{0.4 \cdot 5.67}{401 \cdot K^3} \cdot \left[(T_S / 100)^4 - 12.202^4 \cdot K^4 \right] = \mathbf{8.51 \, K}$$

Discussion

This example demonstrates that a temperature measurement can be distorted by thermal radiation. Particularly large errors can occur in the flow of gases. At high temperatures the deviation can be extremely large as the difference $T_1^4 - T_2^4$ also at smaller deviations has reasonable values. In our case with shield, the temperature difference between sensor and shield is only 27 K but the error is 8.51 K. The high difference is due to the large temperature of 950 °C. At a gas temperature of 100 °C with the shield the error would only be 0.298 K.

To avoid wrong measurements, glass thermometers for room temperature measurements have a polished metal shield installed around the sensor.

7.3   Thermal radiation of gases

Like solids and liquids, some gases have the ability to emit and absorb thermal radiation. Gases constituting only one atom (noble gases) or two atoms per molecule (elementary gases O_2, N_2, H_2 or gases as CO and HCl) are diatherm, i.e., they are transparent for the thermal radiation. Other gases and vapors, consisting of more than two atoms per molecule, as e.g. H_2O, CO_2, SO_2, NH_3 and CH_4 are potential radiators, which emit and absorb thermal radiation in a small range of wavelength. The intensity of thermal radiation of hydrocarbons increases with the number of atom per molecule. For technical application, air can be assumed to be diatherm as the carbon dioxide concentration is very low. However for our climate, the carbon dioxide concentration in the air is of greatest interest. With increasing of the CO_2 concentration the thermal radiation of the earth surface is absorbed and resubmitted to the earth and creates the so called "green house effect".

The amount of heat absorbed by gas is dependent of the thickness s of the gas layer. The intensity of the thermal radiation through a gas layer is given as:

$$i = i_0 \cdot e^{a \cdot s} \tag{7.26}$$

Therein a is the *absorptivity of gases*. The intensity absorbed from the gas can be given by integration of Equation (7.26):

$$i_\alpha = i_0 - i = i_0 \cdot (1 - e^{a \cdot s}) \tag{7.27}$$

The absorptivity of a gas is defined as:

$$\alpha_g = 1 - e^{a \cdot s} \tag{7.28}$$

It is dependent of temperature and pressure. The space occupied by a gas can be of very complex geometry. Instead of the layer thickness, like the hydraulic diameter, an equivalent thickness s_{gl}, is used, which is determined by gas volume V_g and surface area.

$$s_{gl} = f \cdot \frac{4 \cdot V_g}{A_g} \tag{7.29}$$

The correction function f takes into account the geometry and pressure and has approximately the value of 0.9.

7.3.1 Emissivities of flue gases

Thermal radiation of gases is of great interest for the design of combustion chambers. The gases with thermal radiation potential are water vapor and carbon dioxide. The flue gas in the combustion chamber mainly consists of nitrogen and the part of oxygen, not used for the combustion, as the diatherm component. Depending of the fuel, the concentration of the water vapor and carbon dioxide varies. In the following calculation procedures it is assumed that the gas mixture of the flue gas have a diatherm portion (nitrogen and oxygen) and water vapor and carbon dioxide as potential thermal radiators. In this book only the thermal radiation of dust-free gases is discussed. Corrections for flue gases containing solid particles can be found in VDI Heat Atlas [7.1].

The *emissivity of a flue gas* is:

$$\varepsilon_g = \varepsilon_{H_2O} + \varepsilon_{CO_2} - (\Delta\varepsilon)_g \tag{7.30}$$

The absorptivity is given similarly:

$$\alpha_g = \alpha_{H_2O} + \alpha_{CO_2} - (\Delta\varepsilon)_g \tag{7.31}$$

The correction terms $\Delta\varepsilon$ are given in the diagram in Figure 7.10.

7.3.1.1 Emissivity of water vapor

In the diagrams in Figure 7.7 and 7.8 the emissivities of water vapor and its correction function are given as a function of the partial pressure, the pressure and the temperature. The emissivity of gases is not always the same as the absorptivity. It depends on the wall and gas temperature. If the absolute temperature of the wall T_W is not the same as that of the gas T_g, the following correlation has to be used:

$$\alpha_{gW} = \varepsilon_{gW} \cdot (T_g / T_W)^{0.45} \tag{7.32}$$

Therein α_{gW} is the absorptivity of the vapor at the wall. It can be determined in the diagram in Figure 7.7. It is important that the partial pressure has to be determined with the wall temperature.

$$p_{H_2O,W} = p_{H_2O} \cdot (T_W / T_g) \tag{7.33}$$

7.3.1.2   Emissivity of carbon dioxide

In Figure 7.9 the emission coefficients of carbon dioxide are given as a function of the partial pressure and temperature. As with the water vapor the influence of the wall temperature must be taken into account.

$$\alpha_{gW} = \varepsilon_{gW} = (T_g / T_W)^{0.65} \tag{7.34}$$

Therein α_{gW} is the absorptivity of the gas at the wall. In Figure 7.9 the emissivity ε_{gW} is given. Similarly with the water vapor the partial pressure at the wall has to be used.

$$p_{CO_2,W} = p_{CO_2} \cdot (T_W / T_g) \tag{7.35}$$

7.3.2   Heat transfer between gas and wall

The heat rate between gas volume and wall, which envelops this volume, is given as:

$$\dot{Q}_{gW} = \frac{\varepsilon_W \cdot C_s \cdot A}{1 - (1 - \varepsilon_W) \cdot (1 - \alpha_{gW})} \cdot \left[\varepsilon_g \cdot \left(\frac{T_g}{100} \right)^4 - \alpha_{gW} \cdot \left(\frac{T_W}{100} \right)^4 \right] \tag{7.36}$$

In most cases the wall temperature is much lower than that of the gas, therefore the contribution of the term with the lower temperature has only marginal influence, for example, at 400 K wall and 1200 K gas temperature it is only 1.2 %. In such a case with neglecting the influence of the lower temperature, the following approach can be applied:

$$\dot{Q}_{gW} \approx \frac{\varepsilon_g \cdot \varepsilon_W \cdot C_s \cdot A}{1 - (1 - \varepsilon_W) \cdot (1 - \alpha_{gW})} \cdot \left(\frac{T_g}{100} \right)^4 \tag{7.37}$$

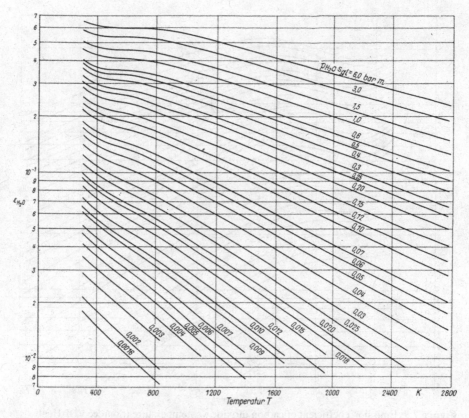

Figure 7.7: Emissivity of water vapor vs. temperature (Source: VDI-Heat Atlas)

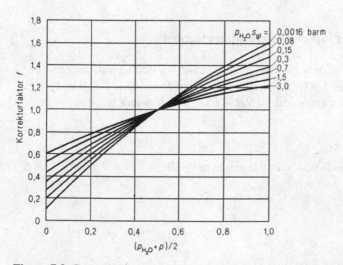

Figure 7.8: Correction function (Korrekturfaktor f) for water vapor vs. partial pressure (Source: VDI-Heat Atlas)

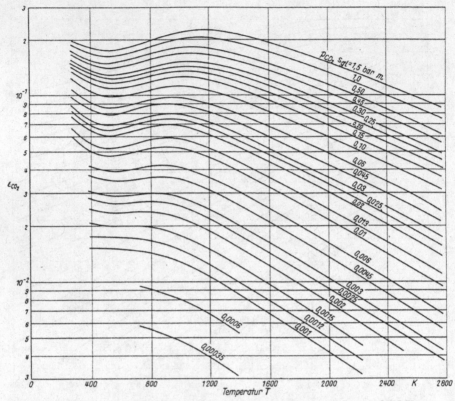

Figure 7.9: Emission coefficient of carbon dioxide vs. temperature (Source: VDI-Heat Atlas)

EXAMPLE 7.5: Heat rate of a combustion chamber

A cubical combustion chamber has an edge length of 0.5 m. The wall temperature is 600 °C, that of the gas 1 400 °C. The emissivity of the wall is 0.9. The flue gas contains 12 Vol% water vapor and 10 Vol% CO_2. The pressure is 1 bar.

Find

The heat flow rate transmitted from the gas to the wall.

Solutions

Assumptions

- The flue gas is homogeneous.
- The temperature in the combustion chamber is constant.

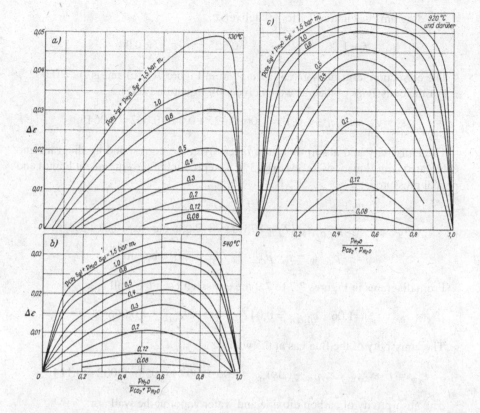

Figure 7.10: Correction terms $\Delta\varepsilon$ for gas mixtures containing water vapor and carbon dioxide a) 130 °C, b) 540 °C, c) 920 °C and over (Source: VDI Heat Atlas)

Analysis

The heat flow rate is calculated with Equation (7.38). First the equivalent length, the emissivity and the absorptivity must be determined. The equivalent length is given by Equation (7.29):

$$s_{gl} = f \cdot \frac{4 \cdot V_g}{A_g} = 0.9 \cdot \frac{4 \cdot a^2}{6 \cdot a} = 0.6 \cdot a = 0.3 \ \text{m}$$

For the determination of the emissivity of water vapor and carbon dioxide the product of equivalent length and partial pressure is required. The partial pressure is equal to the volume fraction.

$$p_{H_2O} \cdot s_{gl} = 0.12 \cdot 0.3 = 0.036 \ \text{bar} \cdot \text{m} \qquad p_{CO_2} \cdot s_{gl} = 0.1 \cdot 0.3 = 0.03 \ \text{bar} \cdot \text{m}$$

The diagrams in Figures 7.7 to 7.9 deliver:

$$f = 1.06 \qquad \varepsilon_{H_2O} = 0.025 \qquad \varepsilon_{CO_2} = 0.04$$

The correction term $\Delta\varepsilon$ is given in diagram in Figure 7.10 c) and is $\Delta\varepsilon = 0.002$. For the emissivity of the gas we receive with Equation (7.31):

$$\varepsilon_g = \varepsilon_{H_2O} + \varepsilon_{CO_2} - (\Delta\varepsilon)_g = 1.06 \cdot 0.025 + 0.04 - 0.002 = 0.0645$$

The emissivity and absorptivity are taken from the corresponding diagram and calculated with Equations (7.34) to (7.37). The product of the equivalent length and partial pressure is:

$$s_{gl} \cdot p_{H_2O,W} = s_{gl} \cdot p_{H_2O} \cdot T_W / T_g = 873/1\,673 \cdot 0.036 \text{ bar} \cdot \text{m} = 0.019 \text{ bar} \cdot \text{m}$$

$$s_{gl} \cdot p_{CO_2,W} = s_{gl} \cdot p_{CO_2} \cdot T_W / T_g = 0.016 \text{ bar} \cdot \text{m}$$

From diagrams in Figures 7.7 to 7.9 the emissivities at the wall are:

$$f = 1.06 \quad \varepsilon_{H_2O,W} = 0.047 \qquad \varepsilon_{CO_2,W} = 0.066$$

The emissivity of the flue gas at the wall is:

$$\varepsilon_{gW} = f \cdot \varepsilon_{H_2O,W} + \varepsilon_{CO_2,W} - (\Delta\varepsilon)_{gW} = 1.06 \cdot 0.067 + 0.058 - 0.002 = 0.114$$

The absorptivity of carbon dioxide and water vapor at the wall are:

$$\alpha_{H_2O,W} = \varepsilon_{H_2O,W} \cdot (T_g / T_W)^{0.45} = 0.063 \qquad \alpha_{CO_2,W} = \varepsilon_{CO_2,W} \cdot (T_g / T_W)^{0.65} = 0.101$$

The absorptivity of the flue gas at the wall is:

$$\alpha_{gW} = \alpha_{H_2O,W} + \alpha_{CO_2,W} - (\Delta\varepsilon)_{gW} = 0.063 + 0.101 - 0.002 = 0.161$$

With Equation (7.36) we receive the heat rate.

$$\dot{Q}_{gW} = \frac{\varepsilon_W \cdot C_s \cdot A}{1 - (1 - \varepsilon_W) \cdot (1 - \alpha_{gW})} \cdot \left[\varepsilon_g \cdot \left(\frac{T_g}{100} \right)^4 - \alpha_{gW} \cdot \left(\frac{T_W}{100} \right)^4 \right] =$$

$$= \frac{0.9 \cdot 5.67 \cdot \text{W} \cdot 6 \cdot 0.5^2 \cdot \text{m}^2}{1 - (1 - 0.9) \cdot (1 - 0.162) \cdot \text{m}^2 \cdot \text{K}^4} \cdot \left[0.0645 \cdot 16.73^4 - 0.162 \cdot 8.73^4 \right] \cdot \text{K}^4 = \mathbf{34.38 \ kW}$$

Discussion

In this example it was assumed that the temperature of the flue gas is constant. Air is blown into the combustion chamber with a fan and together with the fuel during the combustion process produces the flue gas. Due to the thermal radiation but also by forced convection the temperature of the gas flow decreases. In the area of the flames the major part of the heat transfer occurs by thermal radiation. Furthermore, the flue gas can contain soot which improves the heat transfer, but has not been taken into account. The exact calculation of combustion chambers is only possible with 3D computer programs. The calculation procedure given here can serve as a rough estimation only.

8 Heat exchangers

When calculating heat exchangers the analysis can have completely different goals:

- Design of heat exchangers: mass flow rates and temperatures of the fluids are defined, the dimensions of the heat exchanger must be calculated
- Recalculation of heat exchangers: the geometry of the heat exchanger is known and its performance has to be calculated in off-design conditions
- Optimization of heat exchangers and systems
- Mechanical design

In practice the thermal and mechanical design go hand in hand with the goal of optimizing the heat exchanger. In this book only the thermal design is explained, as this is the foundation for all other analyses.

The *heat exchangers* discussed up to here were either parallel-flow or counter-flow heat exchangers or apparatus with constant temperature in one of the fluids (condensation and evaporation). The design of heat exchangers in which the fluid flows in cross-flow cannot be handled with the knowledge acquired up to here.

8.1 Definitions and basic equations

Figure 8.1 shows a schematic of a heat exchanger with important information of the fluid flows [8.1].

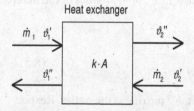

Figure 8.1: Schematic of a heat exchanger

The fluid 1 enters the heat exchanger with the temperature ϑ'_1 and leaves it with the temperature ϑ''_1. The fluid 2 enters the heat exchanger with temperature ϑ'_2 and leaves it with the temperature ϑ''_2. The temperature change of each fluid depends on the mass flow rates, inlet temperatures, specific heat capacities and overall heat

215

transfer coefficient as well as of the exchanger surface area. The heat rate transferred from one to the other fluid is:

$$\dot{Q}_{12} = k \cdot A \cdot \Delta \vartheta_m \qquad (8.1)$$

In equation (8.1) it is assumed that the heat exchanger has a constant mean overall heat transfer coefficient. In most cases this is rather close to reality as usually the mean heat transfer coefficients of the fluids are determined.

In cases where the heat transfer mode changes, e.g. a vapor condenses completely in a tube and then the condensate is sub-cooled, the calculation must be performed with local heat transfer coefficients in the subsequent tube sections.

The log mean temperature difference $\Delta\vartheta_m$ is:

$$\Delta \vartheta_m = \frac{1}{A} \cdot \int_A (\vartheta_1 - \vartheta_2) \cdot dA \qquad (8.2)$$

The local temperature difference of the fluids 1 and 2 is $\vartheta_1 - \vartheta_2$. For parallel-flow or counterflow heat exchangers or in apparatus in which the temperature of one fluid is constant, the log mean temperature difference was defined in Chapter 1.1 as:

$$\Delta \vartheta_m = \frac{\Delta \vartheta_{gr} - \Delta \vartheta_{kl}}{\ln(\Delta \vartheta_{gr} / \Delta \vartheta_{kl})} \qquad \text{for} \quad \Delta \vartheta_{gr} - \Delta \vartheta_{kl} \neq 0 \qquad (8.3)$$

$$\Delta \vartheta_m = (\Delta \vartheta_{gr} + \Delta \vartheta_{kl}) / 2 \qquad \text{for} \quad \Delta \vartheta_{gr} = \Delta \vartheta_{kl} \qquad (8.4)$$

The temperature difference of the fluid flows at the inlet and outlet are $\Delta\vartheta_{gr}$ and $\Delta\vartheta_{kl}$, whereas $\Delta\vartheta_{gr}$ is the lager and $\Delta\vartheta_{kl}$ the smaller difference. In earlier days the validity of Equation (8.4) was limited to $\Delta\vartheta_{gr} - \Delta\vartheta_{kl} < 1$ K. For a heat exchanger with small temperature differences this could result in severe error. Today calculators and computers allow the calculation of the log mean temperature difference even at very small differences (e.g. $\Delta\vartheta_{gr} - \Delta\vartheta_{kl} = 0.00001$ K).

The energy balance equation gives us the heat flow rate as the product of the mass flow rate and the change of the fluid enthalpy.

$$\dot{Q} = \dot{m}_1 \cdot (h_{11} - h_{12})$$
$$\dot{Q} = -\dot{m}_2 \cdot (h_{21} - h_{22}) \qquad (8.5)$$

The enthalpy h_{11} is that of the fluids 1 at the inlet, h_{12} the one at the outlet. Respectively the enthalpy h_{21} is that of the fluids 2 at the inlet and h_{22} the one at the outlet. Equation (8.5) has universal validity, i.e. also for flows with phase change. At flows without phase change the enthalpy can be given as a function of the temperature.

$$\dot{Q} = \dot{m}_1 \cdot c_{p1} \cdot (\vartheta_1' - \vartheta_1'')$$
$$\dot{Q} = -\dot{m}_2 \cdot c_{p2} \cdot (\vartheta_2' - \vartheta_2'') \tag{8.6}$$

In Equation (8.5) heat losses, change of kinetic and potential energy were not taken into consideration. In cases when these energies have a significant influence, what only seldom happens, the corresponding equations of the first law of thermodynamics have to be applied.

In flows without phase change the parameter *heat capacity rate* $\dot{W}$ can be defined.

$$\dot{W}_1 = \dot{m}_1 \cdot (h_{11} - h_{12}) / (\vartheta_1' - \vartheta_1'') = \dot{m}_1 \cdot c_{p1} \dot{W}$$
$$\dot{W}_2 = \dot{m}_2 \cdot (h_{21} - h_{22}) / (\vartheta_2' - \vartheta_2'') = \dot{m}_2 \cdot c_{p2} \tag{8.7}$$

To create generally valid correlations for heat exchangers, the following dimensionless parameters are defined:

Dimensionless log mean temperature difference:

$$\Theta = \frac{\Delta\vartheta_m}{\vartheta_1' - \vartheta_2'} \tag{8.8}$$

Dimensionless temperature changes of the fluid flows 1 and 2 are P_1 and P_2, which are called *cell efficiency*. They are the temperature change of the fluid flow divided by the inlet temperature difference, which is the largest one.

$$P_1 = \frac{\vartheta_1' - \vartheta_1''}{\vartheta_1' - \vartheta_2'} \qquad\qquad P_2 = \frac{\vartheta_2'' - \vartheta_2'}{\vartheta_1' - \vartheta_2'} \tag{8.9}$$

Here the temperatures should be selected so that P_1 is positive, with which also P_2 will be positive.

Number of transfer units NTU of both fluid flows:

$$NTU_1 = \frac{k \cdot A}{\dot{W}_1} = \frac{\vartheta_1' - \vartheta_1''}{\Delta\vartheta_m} \qquad\qquad NTU_2 = \frac{k \cdot A}{\dot{W}_2} = \frac{\vartheta_2'' - \vartheta_2'}{\Delta\vartheta_m} \tag{8.10}$$

The number of transfer units is the ratio of temperature change of the fluid flow to the log mean temperature difference or the ratio of the product of overall heat transfer coefficient and heat transfer area to the heat capacity rate.

Heat capacity rate ratio of the two fluid flows:

$$R_1 = \frac{\dot{W}_1}{\dot{W}_2} = \frac{1}{R_2} \tag{8.11}$$

Within these dimensionless parameters the following relationships exists:

$$\frac{P_1}{P_2} = \frac{NTU_1}{NTU_2} = \frac{1}{R_1} = R_2 \tag{8.12}$$

$$\Theta = \frac{P_1}{NTU_1} = \frac{P_2}{NTU_2} \tag{8.13}$$

8.2   Calculation concepts

There exist a large number of calculation procedures for heat exchangers, which differ in the area of application, calculation effort and accuracy. The most exact procedures are the difference methods and step-by-step procedures. Here the heat exchanger is calculated in segments with local values, e.g., the local flow conditions and temperatures are applied for the calculation of the heat transfer coefficients. With regard to the calculation effort this is the most expensive method. This calculation procedure will not be discussed in this book.

8.2.1   Cell method

In the *cell method* the heat transfer surface area of the heat exchanger is divided into cells, in which the flows are directed in series or in reverse order with full or partial flow rate of the fluids. Each heat transfer surface area has an individual inlet and outlet temperature. Each cell will be applied with most realistic flow directions. Instead of a complete heat exchanger, a system with interconnected individual heat exchangers is created [8.2, 8.3].

With the equations, valid for the flow conditions of the cell and with the given inlet temperatures, the outlet temperatures of each cell can be calculated. For the calculation the term $k \cdot A$ must be known. The determination of $k \cdot A$ takes place with corresponding correlations for the heat transfer coefficients.

Starting with the given inlet temperatures of the fluid flow we receive $2n$ equations for n cells for the $2n$ unknown outlet temperatures. This equation matrix delivers all the intermediate temperatures of both fluid flows. With these temperatures the material properties and heat transfer coefficients are determined. Each individual cell can have different surface areas and heat transfer coefficients. At not too large temperature differences and flow conditions (e.g. flow in a tube and cross flow outside), a constant overall heat transfer for all cells can be assumed. Often heat exchangers are designed such as the overall heat transfer coefficients and the heat transfer areas can be selected equal for each cell. In this case a constant value of $k \cdot A$ allows a simplified calculation and the NTU values for each fluid flow are also constant. The number of internal passes in a heat exchanger is z and that of the external flows n. The number of flows is the number of redirections plus 1.

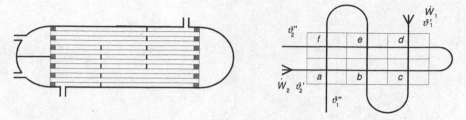

Figure 8.2: Left: heat exchanger with two internal and three external passes; right: cell model

The cell model will be demonstrated here with a heat exchanger having two internal ($n = 2$) and tree external passes ($z = 3$), as shown in Figure 8.2. To simplify the calculation it is assumed that $k \cdot A$ in all cells has the same value and is constant. The heat exchanger consists of six cells with different flow conditions that are signed with index letters. For the whole apparatus the following relations apply:

$$NTU_{1ges} = \frac{k \cdot A}{c_{p1} \cdot \dot{m}_1} \qquad NTU_{2ges} = \frac{k \cdot A}{c_{p2} \cdot \dot{m}_2} = R_1 \cdot NTU_{1ges} \qquad (8.14)$$

The value of $k \cdot A$ is constant and each cell has the same heat transfer area, thus the following relation is valid for each cell:

$$NTU_1 = \frac{k \cdot A_i}{c_{p1} \cdot \dot{m}_1} = \frac{NTU_{1ges}}{n \cdot z}$$

$$NTU_2 = \frac{k \cdot A_i}{c_{p2} \cdot \dot{m}_2} = \frac{NTU_{2ges}}{n \cdot z} = R_1 \cdot NTU_1 \qquad (8.15)$$

The flow conditions of the cell (e.g. pure cross-flow or cross-flow mixed with parallel-flow and tube flow) determine the dimensionless temperatures of the cell. For any cell j, the dimensionless temperatures are defined as:

$$T_{1j} = \frac{\vartheta_{1j} - \vartheta_2'}{\vartheta_1' - \vartheta_2'} \qquad T_{2j} = \frac{\vartheta_{2j} - \vartheta_2'}{\vartheta_1' - \vartheta_2'} \qquad (8.16)$$

The cell j is entered with the fluid flow 1 from cell p and with fluid flow 2 from cell q (Figure 8.3).

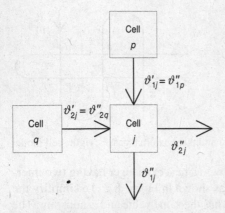

Figure 8.3: Temperatures of cell *j*

For the dimensionless temperatures we receive the following relationships:

$$(1-P_{1j}) \cdot T_{1p}'' - T_{1j}'' + P_{1j} \cdot T_{2q}'' = 0$$
$$(1-P_{2j}) \cdot T_{2q}'' - T_{2j}'' + P_{2j} \cdot T_{1p}'' = 0 \qquad (8.17)$$

For P_{1j} and P_{2j} the inlet respectively outlet temperature has to be inserted in to Equation (8.7).

$$P_{1j} = \frac{\vartheta_{1j}' - \vartheta_{1j}''}{\vartheta_{1j}' - \vartheta_{2j}'} \qquad\qquad P_{2j} = \frac{\vartheta_{2j}'' - \vartheta_{2j}'}{\vartheta_{1j}' - \vartheta_{2j}'} \qquad (8.18)$$

If cell *j* is the inlet cell of the fluid flow 1 or 2, the following applies:

$$T_{1p}'' = T_{1j}' = 1 \qquad\qquad T_{2q}'' = T_{2j}' = 0 \qquad (8.19)$$

Correspondingly if cell *j* is the outlet cell of fluid flow 1 or 2:

$$P_{1ges} = 1 - T_{1j}'' \qquad\qquad P_{2ges} = T_{2j}'' \qquad (8.20)$$

For the example discussed here, we receive the following equations:

$$T_{2a}'' = P_{2a} \cdot T_{1f}'' \qquad\qquad T_{1a}'' = (1-P_{1a}) \cdot T_{1f}'' + P_{1a} \cdot T_{2a}'' = 1 - P_{1ges}$$
$$T_{2b}'' = (1-P_{2b}) \cdot T_{2a}'' + P_{2b} \cdot T_{1c}'' \qquad\qquad T_{1b}'' = (1-P_{1b}) \cdot T_{1c}'' + P_{1b} \cdot T_{2a}''$$
$$T_{2c}'' = (1-P_{2c}) \cdot T_{2b}'' + P_{2c} \cdot T_{1d}'' \qquad\qquad T_{1c}'' = (1-P_{1c}) \cdot T_{1d}'' + P_{1c} \cdot T_{2b}''$$
$$T_{2d}'' = (1-P_{2d}) \cdot T_{2c}'' + P_{2d} \qquad\qquad T_{1d}'' = (1-P_{1d}) + P_{1d} \cdot T_{2c}''$$
$$T_{2e}'' = (1-P_{2e}) \cdot T_{2d}'' + P_{2e} \cdot T_{1b}'' \qquad\qquad T_{1e}'' = (1-P_{1e}) \cdot T_{1b}'' + P_{1e} \cdot T_{2d}''$$
$$T_{2f}'' = (1-P_{2f}) \cdot T_{2e}'' + P_{2f} \cdot T_{1e}'' = P_{2ges} \qquad\qquad T_{1f}'' = (1-P_{1f}) \cdot T_{1e}'' + P_{1f} \cdot T''$$

We do now have 12 equations with 12 unknown temperatures. The equations can be solved with the corresponding mathematical methods. As long as k respectively kA have the same value for all cells and the parameter P_1 and P_2 are equal in each cell, the solution is rather simple. In cases of individual values of $k \cdot A$ of each cell, they must be calculated separately.

Relations for P_i for some flow arrangements are given here.

pure counterflow

$$P_i = \frac{1 - \exp\left[(R_i - 1) \cdot NTU_i\right]}{1 - R_i \cdot \exp\left[(R_i - 1) \cdot NTU_i\right]} \quad \text{for} \quad R_1 \neq 1$$

$$P_1 = P_2 = \frac{NTU}{1 + NTU} \quad \text{for} \quad R_1 = 1 \tag{8.21}$$

pure parallel-flow

$$P_i = \frac{1 - \exp\left[-(R_i + 1) \cdot NTU_i\right]}{1 + R_i} \tag{8.22}$$

pure cross-flow

$$\left\{\begin{array}{l}\left[1 - e^{-NTU_i} \cdot \sum_{j=0}^{m} \frac{1}{j!} \cdot NTU_i^{\,j}\right] \cdot \\ \cdot \left[1 - e^{-R_i \cdot NTU_i} \cdot \sum_{j=0}^{m} \frac{1}{j!} \cdot NTU_i^{\,j}\right]\end{array}\right\} i = 1,2 \tag{8.23}$$

cross-flow to one tube row $\quad P_1 = 1 - \exp\left[(e^{-R_1 \cdot NTU_1} - 1)/R_1\right] \tag{8.24}$

The analysis with the cell method usually requires computer programs. The simpler procedure is the mean log temperature difference method, which will be discussed in Chapter 8.2.2.

EXAMPLE 8.1: Analysis of a heat exchanger with the cell method

The flow arrangement of the heat exchanger consists of two inner and two outer flows with a flow reversal on the shell side. The parameter $k \cdot A$ controlling the heat transfer has for all cells the same value of 4 000 W/K. The shell side flow has the index 1. To simplify the calculation both heat capacity rates $\dot{W}_1$ and $\dot{W}_2$ were selected with 3 500 W/K. The inlet temperature of flow 1 is $\vartheta'_1 = 100\ °C$ and that of the flow 2 $\vartheta'_2 = 20\ °C$.

Find

The outlet temperatures ϑ''_1 and ϑ''_2.

Solution

Schematic   See sketch.

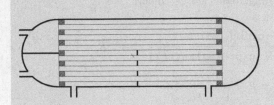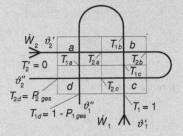

Assumption

• The overall heat transfer coefficient and the heat transfer surface area has in all cells the same value.

Analysis

The dimensionless parameters required for the whole system are:

$$R_1 = R_2 = \dot{W}_1 / \dot{W}_2 = 1$$

$$NTU_{1\,ges} = k \cdot A / \dot{W}_1 = 4\,000 / 3500 = 1.1429 = NTU_{2\,ges}$$

For each cell we receive:

$$NTU_{1,2} = 0.25 \cdot NTU_{1\,ges} = 0.2857$$

As we have here a heat exchanger with one tube row in cross flow, P_1 and P_2 can be determined with Equation (8.24).

$$P_1 = P_2 = 1 - \exp[(e^{-R_1 \cdot NTU_1} - 1) / R_1] = 1 - \exp[(e^{-0.2857} - 1)] = 0.220$$

The temperature changes can be calculated with Equations (8.17) to (8.20). We receive for the eight unknowns eight independent equations. As the parameters P_1 and P_2 have the same value for all cells, the calculation simplifies. Further $P_{1\,ges}$ has the same value as $P_{2\,ges}$. With Equation (8.20) we receive: $P_{1\,ges} + P_{2\,ges} = 1$ and subsequently $P_{1\,ges} = P_{2\,ges} = 0.5$.

$$T_{2a} = P_1 \cdot T_{1b} \qquad\qquad T_{1a} = (1 - P_1) \cdot T_{1b}$$
$$T_{2b} = (1 - P_2) \cdot T_{2a} + P_1 \cdot T_{1c} \qquad T_{1b} = (1 - P_1) \cdot T_{1c} + P_1 \cdot T_{2a}$$
$$T_{2c} = (1 - P_2) \cdot T_{2b} + P_1 \qquad\qquad T_{1c} = (1 - P_1) \cdot 1 + P_1 \cdot T_{2b}$$
$$T_{2d} = P_{2\,ges} \qquad\qquad\qquad T_{1d} = 1 - P_{1\,ges}$$

These eight linear equations can be solved and the dimensionless temperatures can be transformed in Celsius-temperatures. The results are tabulated below:

$$T_{1j} = \frac{\vartheta_{1j} - \vartheta_2'}{\vartheta_1' - \vartheta_2'} = \frac{\vartheta_{1j} - 20\ ^\circ C}{80\ K} \qquad \text{and} \qquad T_{2j} = \frac{\vartheta_{2j} - \vartheta_2'}{\vartheta_1' - \vartheta_2'} = \frac{\vartheta_{2j} - 20\ ^\circ C}{80\ K}$$

T_{2a}	T_{2b}	T_{2c}	T_{2d}	T_{1a}	T_{1b}	T_{1c}	T_{1d}	
0.153	0.306	0.458	0.5	0.542	0.694	0.847	0.5	
ϑ_{2a}	ϑ_{2b}	ϑ_{2c}	ϑ_{2d}	ϑ_{1a}	ϑ_{1b}	ϑ_{1c}	ϑ_{1d}	
32.2	44.5	56.6	60	63.4	75.5	87.8	60.0	°C

The total temperature changes $P_{1\ ges}$ and $P_{2\ ges}$ are equal at 0.5, i.e. the outlet temperatures have the same value of 60 °C.

Discussion

The cell method requires computer codes to solve the equations. Already the simple example here needs a rather high calculation effort. However, some conclusions can be drawn that would not be possible with other methods.

So the flow configuration in this example is not economical, because in cell d all temperatures are close to 60 °C and therefore the heat rate there is very low. A better result would be achieved if cell d were the inlet cell for fluid flow 1.

8.2.2   Analysis with the log mean temperature method

The heat rate can be determined with Equation (8.1). The log mean temperature is only known for parallel-flow, counterflow or when the temperature of one fluid is constant. With the cell method the log mean temperatures can be determined for complex flow arrangements. With the diagrams in Figures 8.5 to 8.12 the log mean temperatures of several flow arrangements can be determined. With the dimensionless temperatures P_1 and P_2 the number of transfer units NTU_1 and NTU_2 can be determined. Equation (8.13) delivers the dimensionless temperature Θ and with Equation (8.8) the log mean temperature difference $\Delta\vartheta_m$ can be calculated.

The diagrams have been proposed by Roetzel and Spang following a review of all known ways to represent the relationship between P_i, NTU_i and $\Delta\vartheta_m$. The diagrams 8.5 to 8.12 have been published in the VDI Heat Atlas and have been taken from there.

Figure 8.4 demonstrates the use of the diagrams. We find the dimensionless temperatures P_1 and P_2 on the bottom and left axis and on the top and right axis the dimensionless ratio of heat capacity flow R_1 and R_2. Here we have to pay attention that the lower value of R is used, for example, its value must be smaller than 1.

In the diagram there are two sets of curves. The upper continuous curves are for NTU_1, the lower ones for NTU_2 the dashed curves are for the parameter F, which is defined as:

$$F = \frac{\Delta\vartheta_m}{\Delta\vartheta_{mG}} = \frac{\Delta\vartheta_m \cdot (\Delta\vartheta_{gr} - \Delta\vartheta_{kl})}{\ln(\Delta\vartheta_{gr} / \Delta\vartheta_{kl})} = \frac{NTU_{iG}}{NTU_i} \tag{8.25}$$

The index G is for a counterflow heat exchanger mean temperature difference according to Equation (8.3).

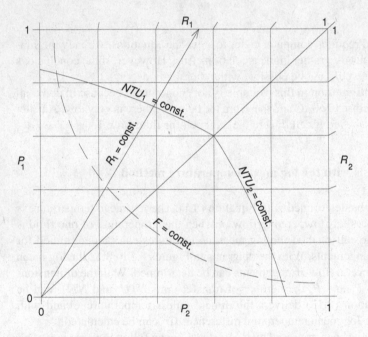

Figure 8.4: How to use the diagrams in Figures 8.5 to 8.12

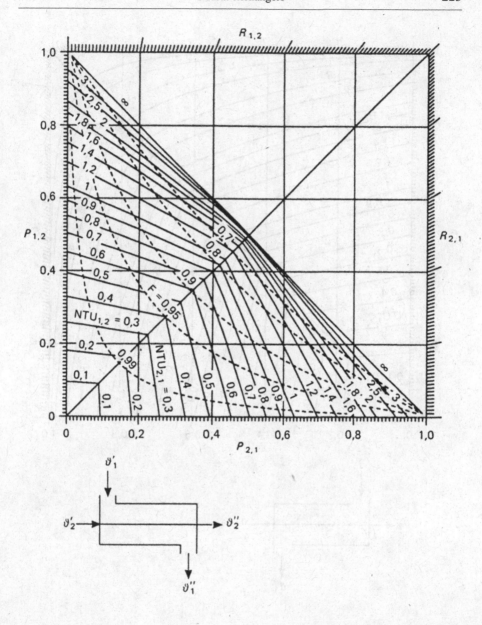

Figure 8.5: Pure parallel-flow (Source: VDI-Wärmeatlas, 9. Aufl.)

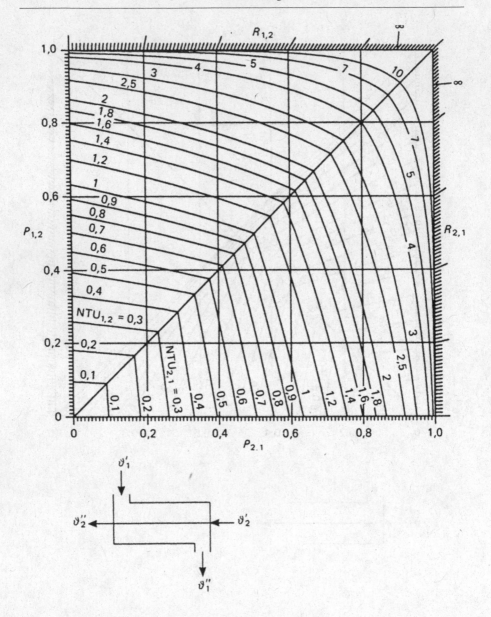

Figure 8.6: Pure couterflow (Source: VDI-Wärmeatlas, 9. Aufl.)

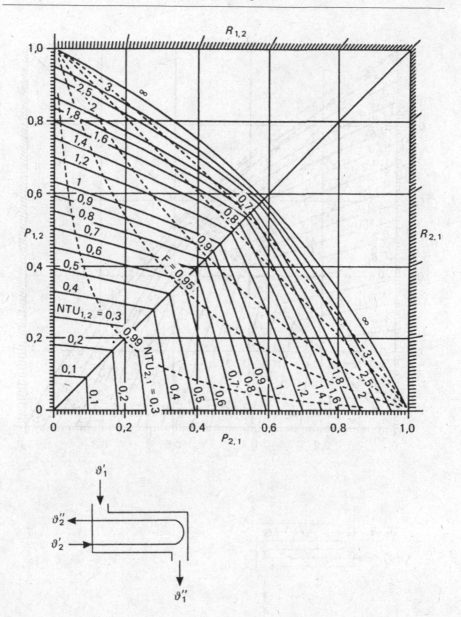

Figure 8.7: Tube and shell heat exchanger with one external and two internal passes (Source: VDI-Wärmeatlas, 9. Aufl.)

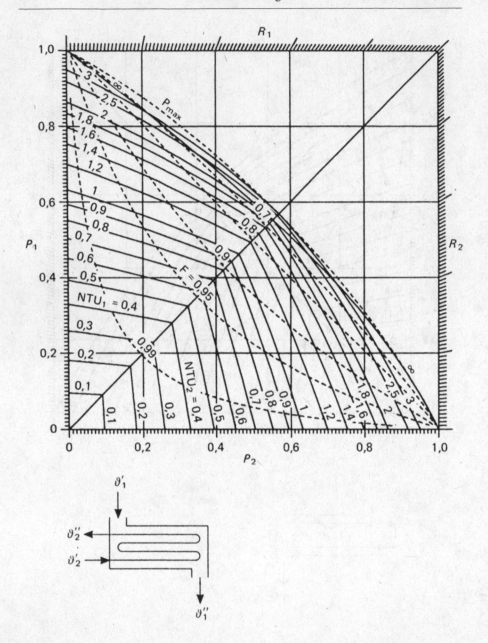

Figure 8.8: Tube and shell heat exchanger with one external and four internal passes
(Source: VDI-Wärmeatlas, 9. Aufl.)

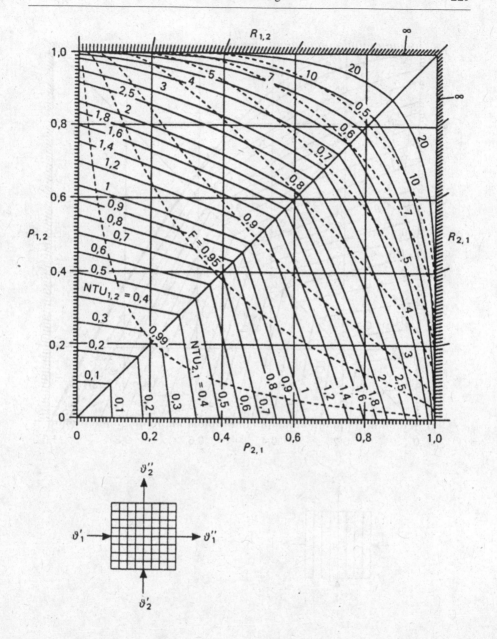

Figure 8.9: Pure cros- flow (Source: VDI-Wärmeatlas, 9. Aufl.)

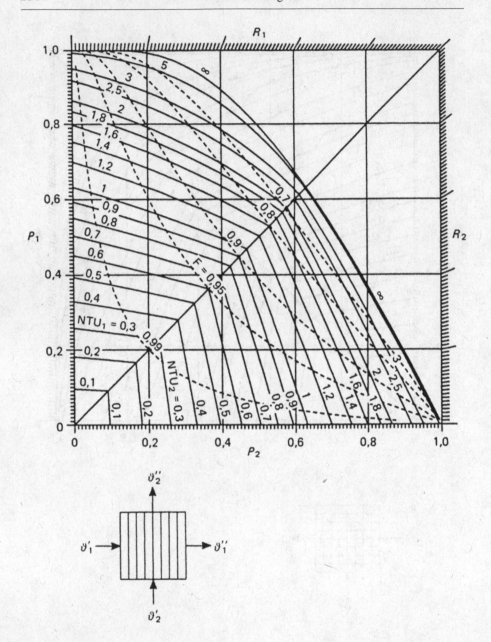

Figure 8.10: Cross-flow with one tube row and cross mixed cross-flow on one side
(Source: VDI-Wärmeatlas, 9. Aufl.)

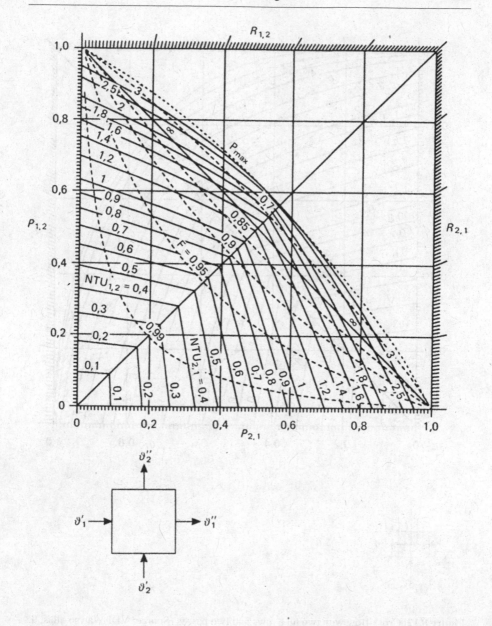

Figure 8.11: Cross-flow cross mixed on both sides (source: VDI-Wärmeatlas, 9. Aufl.)

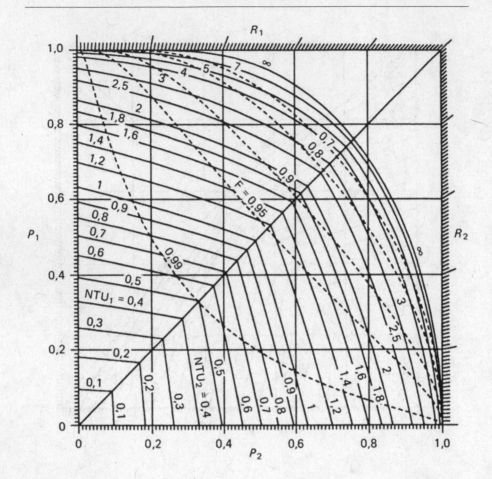

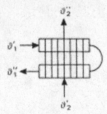

Figure 8.12: Cross-flow with two tube rows and two passes (Source: VDI-Wärme-atlas, 9. Aufl.)

EXAMPLE 8.2: Design of an automotive cooler

The automotive cooler shall transfer 50 kW heat rate at an ambient temperature of 45 °C. The cooling water enters the cooler with a temperature of 94 °C and shall be cooled to 91 °C. The air flows between the fins of the cooler with a velocity of 20 m/s. The square cooling water channels have the following geometry: wall thickness 1 mm, length 550 mm, outside width 6 mm, outside depth 50 mm. On the channels fins with 0.3 mm thickness are soldered in 1 mm distance. The distance between the cooling water channels is 60 mm. The water channel walls and the fins have a thermal conductivity of 120 W/(m K).

The thermal properties of cooling water and air:

Water: $\rho = 963.6$ kg/m³, $\lambda = 0.676$ W/(m K), $\nu = 0.317 \cdot 10^{-6}$ m²/s, $Pr = 1.901$, $c_p = 4.208$ kJ/(kg K).

Air: $\rho = 1.078$ kg/m³, $\lambda = 0.028$ W/(m K), $\nu = 18.25 \cdot 10^{-6}$ m²/s, $Pr = 0.711$, $c_p = 1.008$ kJ/(kg K).

Find

The required number of channels.

Solution

Schematic   See sketch.

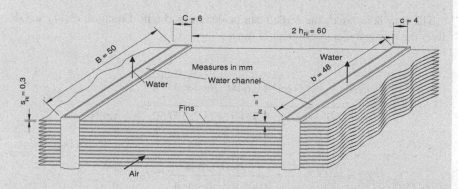

Measures in mm

Assumptions

- In the whole cooler the heat transfer coefficients are constant.
- The influence of the direction of the heat flux can be neglected.

Analysis

The cooling water is signed with the index 1, the air with 2. The mass flow rate of the cooling water can be determined by the energy balance equation:

$$\dot{m}_1 = \frac{\dot{Q}}{c_{p1} \cdot (\vartheta_1' - \vartheta_1'')} = \frac{50 \cdot kW \cdot kg \cdot K}{4.208 \cdot kJ \cdot (94 - 91) \cdot K} = 3.961 \ kg/s$$

The mass flow rate of the air is dependent on the number of cooling water channels. Always one half of a fin, e.g. 30 mm belongs to a cooling water channel. Per cooling water channel the cross-section for the air flow is:

$$A_2 = 2 \cdot h_{Ri} \cdot (t_{Ri} - s_{Ri}) \cdot n = 2 \cdot h_{Ri} \cdot (t_{Ri} - s_{Ri}) \cdot H / t_{Ri} = 2 \cdot h_{Ri} \cdot (1 - s_{Ri} / t_{Ri}) \cdot H =$$
$$= 2 \cdot 0.03 \cdot 0.7 \cdot 0.55 \cdot m^2 = 0.0231 \ m^2$$

With z cooling water channels, we receive for the air mass flow rate:

$$\dot{m}_2 = z \cdot A_2 \cdot \rho_2 \cdot c_2 = z \cdot 0.0231 \cdot m^2 \cdot 1.078 \cdot kg/m^3 \cdot 20 \cdot m/s = z \cdot 0.498 \ kg/s$$

The total mass flow rate of the air increases with number of channels. The cooling water mass flow rate in each channel decreases with the number of channels.

The next step is to determine the heat transfer coefficients.

The air channels between the fins have the following hydraulic diameter:

$$d_{h2} = 4 \cdot h_{Ri} \cdot (t_{Ri} - s_{Ri}) / [2 \cdot (t_{Ri} - s_{Ri} + h_{Ri})] = 2 \cdot 30 \cdot 0.7 \cdot mm^2 / (30.7 \cdot mm) = 1.368 \ mm$$

The *Reynolds* number of the air is:

$$Re_2 = c_2 \cdot d_{h2} / v_2 = 20 \cdot 0.001368 / (18.25 \cdot 10^{-6}) = 1499$$

The flow is laminar; the *Nußelt* can be determined with Equation (3.17) and deliver the heat transfer coefficient of the air.

$$Nu_{d_i, lam} = \sqrt[3]{3.66^3 + 0.644^3 \cdot Pr \cdot (Re_{d_{h2}} \cdot d_{h2} / l)^{3/2}} =$$
$$= \sqrt[3]{3.66^3 + 0.644^3 \cdot 0.711 \cdot (1499 \cdot 1.37 / 50)^{3/2}} = 4.698$$

$$\alpha_2 = \frac{Nu_{d, lam} \cdot \lambda_2}{d_{h2}} = \frac{4.698 \cdot 0.028 \cdot W}{0.001368 \cdot m \cdot m \cdot K} = 96.2 \ \frac{W}{m^2 \cdot K}$$

The heat transfer coefficient of the air is independent of the number of channels, i.e. it remains constant.

In the cooling water channel the hydraulic diameter is:

$$d_{h1} = 4 \cdot b \cdot c / [2 \cdot (b + c)] = 2 \cdot 48 \cdot 4 \cdot mm^2 / (52 \cdot mm) = 7.385 \ mm$$

The total mass flow rate of the cooling water is distributed to the channels and is therefore dependent on the number of channels. The water velocity in the channels:

$$c_1 = \frac{\dot{m}_1}{z \cdot A_1 \cdot \rho_1} = \frac{\dot{m}_1}{z \cdot b \cdot c \cdot \rho_1} = \frac{3.961 \cdot \text{kg} \cdot \text{m}^3}{z \cdot 0.048 \cdot \text{m} \cdot 0.004 \cdot \text{m} \cdot 963.6 \cdot \text{kg} \cdot \text{s}} = \frac{21.421}{z} \frac{\text{m}}{\text{s}}$$

Reynolds number:

$$Re_1 = \frac{c_1 \cdot d_{h1}}{\nu_1} = \frac{21.421 \cdot 0.00738}{z \cdot 0.317 \cdot 10^{-6}} = \frac{499\,015}{z}$$

For less than 50 channels the water flow in the channel remains turbulent. For the clarity of the calculation the simplified Equation (3.14) for the *Nußelt* number is applied.

$$Nu_{d_{h1}} = 0.0235 \cdot (Re_{d_{h1}}^{0.8} - 230) \cdot Pr^{0.48} \cdot [1 + (d_{h1} / l_1)^{2/3}] = 1222,7 \cdot z^{-0.8} - 7.773$$

The from z dependent heat transfer coefficient is:

$$\alpha_1 = Nu_{d_{h1}} \cdot \lambda_1 / d_{h1} = (111\,932 \cdot z^{-0.8} - 711.5) \cdot \text{W} / (\text{m}^2 \cdot \text{K})$$

To receive the overall heat transfer coefficient with Equation (3.46) the fin efficiency as well as the surface areas A, A_0 and A_{Ri} must be determined. Equation (2.58) delivers the fin efficiency.

$$\eta_{Ri} = \frac{\tanh(m \cdot h_{Ri})}{m \cdot h_{Ri}} = 0.444 \quad \text{with} \quad m = \sqrt{\frac{\alpha_2 \cdot U_{Ri}}{\lambda_{Ri} \cdot A_{Ri}}} = \sqrt{\frac{\alpha_2 \cdot 2 \cdot (B + s_{Ri})}{\lambda_{Ri} \cdot B \cdot s_{Ri}}} = \frac{73.31}{\text{m}}$$

$$A = 2 \cdot (C + B) \cdot H = 0.0616 \, \text{m}^2 \qquad A_0 = 2 \cdot B \cdot H \cdot (1 - s_{Ri} / t_{Ri}) = 0.0385 \, \text{m}^2$$

$$A_{Ri} = 2 \cdot h_{Ri} \cdot B \cdot H / t_{Ri} = 1.65 \, \text{m}^2$$

The overall heat transfer coefficient:

$$\frac{1}{k} = \frac{A}{A_0 + A_{Ri} \cdot \eta_{Ri}} \cdot \frac{1}{\alpha_a} + \frac{s_o}{\lambda_R} + \frac{A}{A_i} \cdot \frac{1}{\alpha_i} =$$

$$= \frac{A}{A_0 + A_{Ri} \cdot \eta_{Ri}} \cdot \frac{1}{\alpha_a} + \frac{s_o}{\lambda_R} + \frac{C + B}{c + b} \cdot \frac{1}{\alpha_i} = \left(\frac{1}{1191} + \frac{1}{111\,932 \cdot z^{-0.8} - 711.5} \right) \frac{\text{m}^2 \cdot \text{K}}{\text{W}}$$

The outlet temperature of the air is dependent of the number of channels. It is required to determine the dimensionless temperatures, the ratio of heat capacity rates and the reference log mean temperature of the counterflow heat exchangers. To determine the log mean temperature of the cooler the listed values are shown tabular. With them the value of the parameter F can be taken from diagram in Figure 8.9.

The air outlet temperature is:

$$\vartheta_2'' = \vartheta_2' + \frac{\dot{Q}}{c_{p2} \cdot \dot{m}_2} = 45\ ^\circ\text{C} + \frac{75 \cdot \text{kW} \cdot \text{kg} \cdot \text{K} \cdot \text{s}}{z \cdot 1.008 \cdot \text{kJ} \cdot 0.498 \cdot \text{kg}} = 45\ ^\circ\text{C} + 99.6\ \text{K}\,/\,z$$

The ratio of heat capacity rates are calculated as:

$$R_2 = \frac{\dot{m}_2 \cdot c_{p2}}{\dot{m}_1 \cdot c_{p1}} = \frac{z \cdot 0.498 \cdot 1008}{3.961 \cdot 4208} = 0.03012 \cdot z$$

As long as z remains smaller than 37, the value of R_2 remains also smaller than 1 and we can take the values of F below the diagonal. The dimensionless temperature P_2 can be determined with Equation (8.9).

$$P_2 = \frac{\vartheta_2'' - \vartheta_2'}{\vartheta_1' - \vartheta_2'} = \frac{99.6\ \text{K}}{z \cdot 49\ \text{K}} = \frac{2.033}{z}$$

With an assumed number of channels R_2, P_2, F, the log mean temperature difference, the overall heat transfer coefficient and the required surface area A_{tot} can be determined. The number of channels results as: $z_{requ} = A_{tot}/A$.

$$A_{tot} = \frac{\dot{Q}}{k \cdot \Delta\vartheta_m} = \frac{\dot{Q}}{k \cdot F \cdot \Delta\vartheta_{mG}}$$

z	ϑ_2''	R_2	P_2	F	$\Delta\vartheta_{mG}$	$\Delta\vartheta_m$	k	A_{tot}	z_{requ}
-	°C	-	-	-	K	K	W/(m² K)	m²	-
6	61.6	0.18	0.34	0.992	38.4	38.1	1 135	1.158	19
19	50.2	0.57	0.11	0.996	44.8	44.6	1 054	1.064	17
17	50.9	0.51	0.12	0.996	44.4	44.2	1 065	1.061	17

Discussion

The calculation with the log mean temperature difference is easily possible. The example above could have been calculated with a good accuracy as counterflow heat exchanger as F had a value of almost 1.

8.3   Fouling resistance

Up to here only the resistance of the separating walls of the heat exchanger were taken into account. The walls of a heat exchanger are made of solid materials (mostly metal, plastic, glass, graphite, etc.). The metal surfaces are covered by an oxide layer, whose thermal conductivity is lower than that of the metal. Further on some solid particles or salt in the flowing fluid can build deposits. The formation of a thick oxide layer (rust) is possible. Also the desired formation of a protective layer (steam power plant condenser with brass tubes) is possible. All these layers are thermal resistances.

A tube wall with several additional layers has a larger thermal resistance than the "clean" tube. The resistance caused by additional layers is called *fouling resistance* or *fouling*. Here the influence of the fouling resistance on the overall heat transfer coefficient of a tube will be analyzed. For other geometries, e.g. plane walls similar calculation procedures apply.

The exact determination of the effect of fouling takes into account the influence of the thermal resistance of the fouling layers as well as that of the change of geometry. For this calculation a fouling layer, with given thickness and thermal conductivity, is assumed on both sides of the tube wall. With the fouling layers the inside and outside diameter of the tube is influenced. Figure 8.13 illustrates the calculation model.

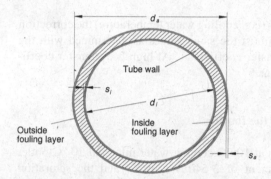

Figure 8.13: Illustration of the exact fouling resistance model

For the fouled overall heat transfer coefficient k_f related to the clean tube outside diameter we receive.

$$\frac{1}{k_f} = \frac{1}{\alpha_a} \cdot \frac{d_a}{d_a + 2 \cdot s_a} + \frac{d_a}{2 \cdot \lambda_{sa}} \cdot \ln\left(\frac{d_a + 2 \cdot s_a}{d_a}\right) + \frac{d_a}{2 \cdot \lambda_R} \cdot \ln\left(\frac{d_a}{d_i}\right) +$$
$$+ \frac{d_a}{2 \cdot \lambda_{si}} \ln\left(\frac{d_i}{d_i - 2 \cdot s_i}\right) + \frac{d_a}{d_i - 2 \cdot s_i} \cdot \frac{1}{\alpha_i} \tag{8.26}$$

Therein are s_a and s_i the thicknesses, λ_{sa} and λ_{si} the thermal conductivities of the outer and inner fouling layer. Mostly the exact thickness and also the thermal conductivity of the fouling layers cannot be determined exactly. From tests the decrease of the overall heat transfer coefficients is known. Based on experience the fouling resistances are given. For some heat exchangers the inner and outer fouling resistances R_i and R_a are known. In most cases only a cumulated fouling resistance R_F is known. With these values we receive the fouled overall heat transfer coefficient.

$$\frac{1}{k_f} = R_a + R_i + \frac{1}{\alpha_a} + \frac{d_a}{2 \cdot \lambda_R} + \frac{d_a}{d_i} \cdot \frac{1}{\alpha_i} = R_F + \frac{1}{\alpha_a} + \frac{d_a}{2 \cdot \lambda_R} + \frac{d_a}{d_i} \cdot \frac{1}{\alpha_i} \quad (8.27)$$

Due to the fouling resistances the overall heat transfer coefficients can strongly be reduced. The reduction increases with the value of the heat transfer coefficients. When designing a heat exchanger the corresponding fouling resistances have to be considered.

Some clients, mainly in the USA, ask instead of the fouling resistance for a *fouling factor* φ, which has to be multiplied with the clean overall heat transfer coefficient.

$$k = k_{clean} \cdot \varphi = \left(\frac{1}{\alpha_a} + \frac{d_a}{2 \cdot \lambda_R} + \frac{d_a}{d_i} \cdot \frac{1}{\alpha_i} \right)^{-1} \cdot \varphi \quad (8.28)$$

With changing of outer parameters (e.g. cooling water temperature) the correction with the fouling factors provides almost the same values as determined with the fouling resistance at lower heat transfer coefficients. At high heat transfer coefficients the deviation can be significant.

EXAMPLE 8.3: Consideration of the fouling

A steam power plant condenser has at 10 °C cooling water inlet and 20 °C outlet temperature a heat transfer coefficient of 3 540 W/(m² K) and the saturation temperature of 25 °C. At a cooling water inlet temperature of 25 °C the outlet temperature is 35 °C. The calculated saturation temperature increases to 39 °C. The fouling resistance has the value of 0.0565 (m² K)/kW.

Find

The expected saturation temperature if instead of the fouling resistance the fouling factor of 0.8 is used.

Assumption

*   The heat rate to the condenser is constant and independent of the cooling water inlet temperature.

Analysis

The heat flux can be determined by the rate equation. For the determination we need the overall heat transfer coefficient at 10 °C cooling water inlet temperature.

$$\Delta\vartheta_m = \frac{\vartheta_2 - \vartheta_1}{\ln\left(\dfrac{\vartheta_s - \vartheta_1}{\vartheta_s - \vartheta_2}\right)} = \frac{10\ \text{K}}{\ln(15/5)} = 9.102\ \text{K}$$

$$\dot{q} = k \cdot \Delta\vartheta_m = 3540 \cdot \text{W/(m}^2 \cdot \text{K)} \cdot 9.102\ \text{K} = 32\,222\ \text{W/m}^2$$

At 25 °C cooling water inlet temperature the log mean temperature difference is:

$$\Delta\vartheta_m = \frac{\vartheta_2 - \vartheta_1}{\ln\left(\dfrac{\vartheta_s - \vartheta_1}{\vartheta_s - \vartheta_2}\right)} = \frac{10\ \text{K}}{\ln(14/4)} = 7.982\ \text{K}$$

This indicates that at the higher cooling water temperature the overall heat transfer coefficient has a higher value. This is due to the changed material properties of the condensate.

$$k = \dot{q} / \Delta\vartheta_m = \frac{32\,225\ \text{W} / \text{m}^2}{7.982\ \text{K}} = 4.037\ \frac{\text{W}}{\text{m}^2\ \text{K}}$$

With the fouling resistances calculated with "clean" overall heat transfer coefficients are:

$$k_{clean,10°C} = (1/k - R_v)^{-1} = (1/3\,540 - 0.0565 \cdot 10^{-3})^{-1} = 4\,425\ \text{W/(m}^2\ \text{K)}$$

$$k_{clean,25°C} = (1/k - R_v)^{-1} = (1/4\,037 - 0.0565 \cdot 10^{-3})^{-1} = 5\,230\ \text{W/(m}^2\ \text{K)}$$

The overall heat transfer coefficient calculated with the fouling factor 0.8 results at 10 °C cooling water inlet temperature in 3 540 W/(m² K) and at 25 °C in a higher value of 4 184 W/(m² K). The difference is 3.51 %. The log mean temperature calculated with this over heat transfer coefficient delivers 7.702 K. For the saturation temperature we receive:

$$\vartheta_s = \frac{\vartheta_1 - \vartheta_2 \cdot \exp[(\vartheta_2 - \vartheta_1)/\Delta\vartheta_m]}{1 - \exp[(\vartheta_2 - \vartheta_1)/\Delta\vartheta_m]} = \frac{25 - 35 \cdot \exp[10/7.702]}{1 - \exp[10/7.702]} = \mathbf{38.8\ °C}$$

Discussion

The heat transfer coefficient determined with the fouling factor is lower than the one determined with the fouling resistance. It pretends a lower condenser pressure. With regard to physics the fouling factor is a wrong approach. The real values are those calculated with the fouling resistance. The condenser designed with the fouling fac-tor would have a 3.5 % too small surface area.

8.4   Tube vibrations

Flow instabilities can lead to critical tube vibrations resulting in high sound levels. The first effect leads to critical oscillations followed by tube destruction, the second to unacceptable noise emission.

8.4.1   Critical tube oscillations

Tube vibrations are induced by flow instability. The vibration can be induced by cross-flow with vortex generation or by resonances of the fluid in the shell and also at parallel-flow between long non-supported tubes. To avoid damages the dangerous oscillations have to be damped. Damping can be achieved by the installation of support plates or support grids. The allowable non-supported length can be determined. It is dependent on the fluid flow velocity, tube material, tube moment of inertia, tube fixation, damping parameters (logarithmic decrement). An exact analysis of the vibration generation modes and the damping parameter is necessary to certainly avoid vibration. Most heat exchangers use an equidistant installation of support plates or grids. This must be a result of an exact analysis. Tubes hit by a fluid jet must be separately investigated. A detailed description of the vibration analysis is published in VDI Heat Atlas, Chapter O [8.5] and in TEMA Standards Of The Tubular Heat Exchangers Manufacturers Association, 9th Edition, Section 6 [8.6]. Here only a simple calculation method is given.

The equation is based on an empiric equation. It allows the determination of the maximum supported tube length l_0, with which dangerous tube vibrations can be avoided. The calculation requires the tube parameters and the flow velocity between the tubes.

$$c_{Sp}^2 \cdot \rho \cdot l_0^5 \cdot \frac{d_a^2}{(d_a^4 - d_i^4) \cdot E} \cdot \frac{10^6}{\text{m}^3} \leq 4.5 \tag{8.29}$$

Therein c_{Sp} is the flow velocity between the tubes, d_a the outside, d_i the inside and E the elasticity module of the tube material. The term $10^6 / \text{m}^3$ was set in to have the result without decimal power and dimensions.

The flow velocity and supported tube length can be influenced by the mechanical design. The support length of the tube can be changed by the number of support structures. Is the result of Equation (8.29) larger then 4.5, it can be reduced by re-ducing the supported tube length. Are the support distances equally long, instead of l_0 the total tube length l_{ges} divided by the number of support places plus 1 $l_0 = l_{ges} /$ (N+1) are inserted and the number of support plates is determined.

EXAMPLE 8.4   Determination of the number of support plates

For the condenser designed in Example 5.3 a vibration analysis shall be performed. The following data can be used:

d_a = 12 mm, d_i = 10 mm, m_{R134a} = 0.5 kg/s, ρ = 66.3 kg/m³, l_{ges}: 5.983 m, n = 47. The elasticity module of copper is 110 kN/mm² = $1.1 \cdot 10^{11}$ N/m². The tube distance is s = 17 mm.

Find

The number of support plates, required to avoid vibration.

Solution

Schematic   See sketch.

Assumption

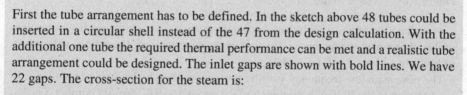

- The flow into the bundle is uniform.

Analysis

First the tube arrangement has to be defined. In the sketch above 48 tubes could be inserted in a circular shell instead of the 47 from the design calculation. With the additional one tube the required thermal performance can be met and a realistic tube arrangement could be designed. The inlet gaps are shown with bold lines. We have 22 gaps. The cross-section for the steam is:

$$A_{in} = 22 \cdot l \cdot s = 22 \cdot 5.983 \text{ m} \cdot 0.005 \text{ m} = 0.658 \text{ m}^2$$

The flow velocity between the tubes we receive as:

$$c_{Sp} = \frac{\dot{m}}{\rho \cdot A_{in}} = \frac{0.5 \cdot \text{kg} \cdot \text{m}^3}{0.658 \cdot \text{m}^2 \cdot 66.3 \cdot \text{kg}} = 0.011 \frac{\text{m}}{\text{s}}$$

Now the vibration criteria can be determined.

$$c_{Sp}^2 \cdot \rho \cdot l_0^5 \cdot \frac{d_a^2}{(d_a^4 - d_i^4) \cdot E} \cdot \frac{10^6}{\text{m}^3} \le 4.5$$

Without support plate the result is 8.1 and for 1 support plate it is 0.25.

Discussion

In this case according to the vibration criteria without a support plate the tubes would be subject to vibration. With one support plate the result is far below the limit. With a supported length of 5.313 m the vibration criteria has the value of 4.5.

8.4.2 Acoustic resonance

With perpendicular cross flow on cylindric bodies above *Reynolds* numbers of 50 *vortex shedding* occurs. On the cylinder surface vortices are generated which shed from the cylinder and remain in the flow. The formation and shedding of the vortices is periodic and occurs alternatively on the lower and upper side of the cylinder. Figure 8.14 illustrates this process. With shedding of the vortices on the back side of the cylinder pressure changes occur, which are forming periodic forces acting on the cylinder.

Figure 8.14: Vortex shedding on a cylinder

The whistling of wires in the wind is generated by vortex shedding. The vibration of the tubes generated by these vortices was discussed above. Here the development of high sound levels will be explained.

The dimensionless frequency ν of the vortex shedding is the *Struhal number S*. It is a function of the Reynolds number and the geometry. The *Struhal* number S is defined as:

$$S = \frac{\nu_{St} \cdot d}{c_0}$$

(8.30)

Therein ν_{St} is the frequency of vortex shedding and is called *Struhal frequency*, d is the cylinder diameter and c_0 the flow velocity. For *Reynolds* numbers less than 10^5 the *Struhal* number of bodies in perpendicular cross flow has a value of approximately 0.2. At higher *Reynolds* numbers the value is between 0.17 and 0.32. Geometry, surface characteristics, in tube bundles the tube arrangement are further influencing variables. In tube bundles, where vortex shedding of one tube influences the frequency of the following tubes, *Struhal* numbers between 1.0 and 1.6 were measured.

In case that the *Struhal* frequency, given by the *Struhal* number, is equal or close to the body's Eigenfrequency, the resonance can leads to so severe oscillations of the tubes and damages may happen. The Tacoma Narrows Bridge, a large bridge of several hundreds of meters span in US state Washington broke in 1940 at wind velocities of 70 km/h due to such strong oscillations.

Every *Struhal* frequency has an accoustic wave length. If flow boundaries, for example, the bundle height, are close to the accoustic wavelength, a so called *lock-in resonance* can occur. This effect must not lead to damage, but sound levels of unacceptable intensity can be generated. The problem with the lock-in resonance is, that even with reasonably low flow velocity the resonance remains.

When designing a heat exchanger the engineer has to make sure that dangerous resonances cannot occur.

EXAMPLE 10.4: Accoustic resonance in a tube bundle

For the reheater bundle in Example 3.6 the *Struhal* number was measures as 1.04. The height of the bundle was 1.4 m but divided in two halves of 675 mm. The tube diameter is 15 mm and the steam velocity 6 m/s. The velocity of sound is 517 m/s.

Find

Has the bundle lock-in resonance and at what velocity lock-in resonance could occur.

Solution

Schematic   See sketch.

Assumption

*   The *Struhal* number is related to the outer diameter and the steam velocity c_0.

Analysis

The accoustic wave length is:

$$\lambda_a = \frac{a}{v}$$

The frequency v to a wavelength of 0.675 m has the value of:

$$v = a / l_a = (517 \text{ m/s}) / 0.765 \text{ m} = 766 \text{ Hz}.$$

With (8.29) the *Struhal* frequency is:

$$v_{St} = \frac{S \cdot c_0}{d} = \frac{1.04 \cdot 6 \cdot \text{m}}{0.015 \cdot \text{m} \cdot \text{s}} = 416 \text{ Hz}$$

The Struhal frequency is much lower than the frequency belonging to a wave length of 0.675 m. Therefore the occurrence of lock-in resonance has not to be anticipated. The steam velocity at which a lock-in resonance my occur is:

$$c_0 = \frac{v_{St} \cdot d}{S} = \frac{765 \cdot s^{-1} \cdot 0.015 \cdot m}{1.04} = \mathbf{11.0 \ m/s}$$

Discussion

With the design of the reheater bundle no lock-in resonance has to be anticipated. Only at a much higher steam velocity of 11 m/s, lock in may happen.

Appendix

A1:   Important physical constants

Critical properties of matter

Material	Chemical formula	Mol mass kg/kmol	T_{crit} K	p_{crit} bar	z_{crit} –
Acetylene	C_2H_2	26.0380	309.0	26.8	0.274
Ammonia	NH_3	17.0305	406.0	112.8	0.284
Argon	Ar	39.9480	151.0	48.6	0.242
Butane	C_4H_{10}	58.1240	425.0	38.0	0.274
Ethane	C_2H_6	30.0700	305.0	48.8	0.285
Ethanol	C_2H_5OH	46.0690	516.0	63.8	0.249
Ethylene	C_2H_4	28.0540	283.0	51.2	0.270
Freon 134a	CF_3CH_2F	102.0300	374.2	40.7	0.260
Helium	He	4.0026	5.2	2.3	0.300
Carbon dioxide	CO_2	44.0100	304.0	73.9	0.276
Carbon monoxide	CO	28.0100	133.0	35.0	0.294
Methane	CH_4	16.0430	191.0	46.4	0.290
Octane	C_8H_{18}	114.2310	569.0	24.9	0.258
Propane	C_3H_8	44.0970	369.8	42.4	0.276
Oxygen	O_2	31.9988	154.0	50.5	0.290
Sulfur dioxide	SO_2	64.0650	431.0	78.7	0.268
Nitrogen	N_2	28.0134	126.0	33.9	0.291
Water	H_2O	18.0153	647.1	220.6	0.233
Hydrogen	H_2	2.0159	33.2	13.0	0.304

Fundamental constants

Avogadro-constant	$N_A = (6.0221367 \pm 0.0000036) \cdot 10^{23}$	mol^{-1}
Universal gas constant	$R_m = (8314.41 \pm 0.07)$	J/(kmol K)
Boltzmann-constant R_m/N_A	$k = (1.380641 \pm 0.000012) \cdot 10^{-23}$	J/K
Elementary charge	$e = (1.60217733 \pm 0.00000049) \cdot 10^{-19}$	C
Planck-constant	$h = (6.6260755 \pm 0.000004) \cdot 10^{-23}$	J s
Speed of light	$c = 299\,792\,458$	m/s
Stefan-Boltzmann-constant	$\sigma_s = 5.670 \cdot 10^{-8}$	W/(m^2 K^4)

Source: [9.3]

A2:   Thermal properties of sub-cooled water at 1 bar pressure

ϑ °C	ρ kg/m³	λ W/(m K)	η 10⁻⁶ kg/(m s)	v 10⁻⁶ m²/s	c_p kJ/(kg K)	Pr -	β 1/K
0	999.8	0.5611	1791.5	1.7918	4.219	13.473	−0.068
2	999.9	0.5649	1673.4	1.6735	4.213	12.480	−0.032
4	1000.0	0.5687	1567.2	1.5673	4.207	11.595	0.001
6	999.9	0.5725	1471.4	1.4715	4.203	10.802	0.032
8	999.9	0.5763	1384.7	1.3849	4.199	10.089	0.061
10	999.7	0.5800	1305.9	1.3063	4.195	9.445	0.088
12	999.5	0.5838	1234.0	1.2346	4.193	8.862	0.114
14	999.2	0.5875	1168.3	1.1692	4.190	8.332	0.139
16	998.9	0.5912	1108.1	1.1092	4.188	7.849	0.163
18	998.6	0.5949	1052.7	1.0541	4.186	7.408	0.185
20	998.2	0.5985	1001.6	1.0034	4.185	7.004	0.207
22	997.8	0.6020	954.4	0.9566	4.183	6.633	0.227
24	997.3	0.6055	910.7	0.9132	4.182	6.291	0.247
26	996.8	0.6089	870.2	0.8730	4.181	5.976	0.266
28	996.2	0.6122	832.5	0.8356	4.181	5.685	0.285
30	995.7	0.6155	797.3	0.8008	4.180	5.415	0.303
32	995.0	0.6187	764.6	0.7684	4.179	5.165	0.320
34	994.4	0.6218	733.9	0.7381	4.179	4.932	0.337
36	993.7	0.6248	705.2	0.7097	4.179	4.716	0.353
38	993.0	0.6278	678.3	0.6831	4.179	4.515	0.369
40	992.2	0.6306	653.0	0.6581	4.179	4.327	0.385
42	991.4	0.6334	629.2	0.6346	4.179	4.151	0.400
44	990.6	0.6361	606.8	0.6125	4.179	3.986	0.415
46	989.8	0.6387	585.7	0.5917	4.179	3.832	0.429
48	988.9	0.6412	565.7	0.5720	4.179	3.687	0.444
50	988.0	0.6436	546.9	0.5535	4.180	3.551	0.457
52	987.1	0.6459	529.0	0.5359	4.180	3.423	0.471
54	986.2	0.6482	512.1	0.5193	4.181	3.303	0.484
56	985.2	0.6503	496.1	0.5035	4.181	3.189	0.498
58	984.2	0.6524	480.9	0.4886	4.182	3.082	0.510
60	983.2	0.6544	466.4	0.4744	4.183	2.981	0.523
62	982.2	0.6563	452.7	0.4609	4.184	2.885	0.536
64	981.1	0.6581	439.6	0.4480	4.185	2.795	0.548
66	980.0	0.6599	427.1	0.4358	4.186	2.709	0.560
68	978.9	0.6616	415.2	0.4242	4.187	2.628	0.572
70	977.8	0.6631	403.9	0.4131	4.188	2.551	0.584
72	976.6	0.6647	393.1	0.4025	4.189	2.478	0.596
74	975.5	0.6661	382.7	0.3924	4.191	2.408	0.607
76	974.3	0.6675	372.9	0.3827	4.192	2.342	0.619
78	973.0	0.6688	363.4	0.3735	4.194	2.279	0.630
80	971.8	0.6700	354.4	0.3646	4.196	2.219	0.642
82	970.5	0.6712	345.7	0.3562	4.197	2.162	0.653
84	969.3	0.6723	337.4	0.3481	4.199	2.107	0.664
86	968.0	0.6734	329.4	0.3403	4.201	2.055	0.675
88	966.7	0.6744	321.8	0.3329	4.203	2.005	0.686
90	965.3	0.6753	314.4	0.3257	4.205	1.958	0.697
92	964.0	0.6762	307.4	0.3188	4.207	1.912	0.708
94	962.6	0.6770	300.6	0.3123	4.209	1.869	0.719
96	961.2	0.6777	294.1	0.3059	4.212	1.827	0.730
98	959.8	0.6784	287.8	0.2998	4.214	1.788	0.740

Source [9.1]

A2:   Thermal properties of sub-cooled water at 1 bar (cont.)

Polynomes for the range of 0 °C to 100 °C.

$$\sum_{i=0}^{6} C_i \cdot \vartheta_R^{9i} \quad \text{with } \vartheta_R = \vartheta / \vartheta_{ref} = \vartheta / 100 \text{ K}$$

	ρ	λ	η	c'_p	Pr	β
C_0	999.85	0.56112	1.79016	4.21895	13.460	−0.06755
C_1	5.4395	0.18825	−6.11398	−0.3299	−51.371	1.8226
C_2	−76.585	0.03255	15.1225	1.15869	134.214	−3.1122
C_3	43.993	−0.23117	−26.7663	−2.35378	−244.081	5.3644
C_4	−14.386	0.15512	30.435	2.88758	282.198	−6.2277
C_5	0	−0.00988	−19.323	−1.84461	−181.31	4.0943
C_6	0	−0.01697	5.1396	0.47973	48.664	−1.1233
Results in	kg/m³	W/(m K)	10^{-3} kg/(m s)	kJ/kg	-	1/K
Std.-Div. %	0.225	0.003	0.004	0.084	0.006	0.130

Important: Not valid for temperatures over 100 °C!

A3: Thermal properties of saturated water and steam

ϑ	p	ρ'	ρ''	λ'	λ''	c'_p	c''_p	η'	η''	ν'	ν''	Pr'	Pr''	r
°C	bar	kg/m³	kg/m³	10⁻³ W/(m K)	10⁻³ W/(m K)	kJ/(kg K)	kJ/(kg K)	10⁻⁶ kg/(m s)	10⁻⁶ kg/(m s)	10⁻⁶ m²/s	10⁻⁶ m²/s	-	-	kJ/kg
0.01	0.00612	999.79	0.00485	561.0	17.1	4.220	1.888	1791.0	9.20	1.7914	1894	13.472	1.017	2500.9
10	0.01228	999.65	0.00941	579.5	17.2	4.196	1.896	1316.7	9.62	1.3171	1023	9.533	1.063	2477.2
20	0.02339	998.16	0.01731	598.0	17.7	4.185	1.906	1007.8	9.89	1.0096	571.2	7.053	1.063	2453.5
30	0.04247	995.61	0.03041	615.2	18.5	4.180	1.918	798.7	10.09	0.8022	331.9	5.427	1.045	2429.8
40	0.07384	992.18	0.05124	630.6	19.4	4.179	1.932	651.9	10.30	0.6570	201.0	4.320	1.024	2406.0
50	0.1235	988.01	0.08314	643.7	20.4	4.180	1.948	545.2	10.53	0.5518	126.7	3.540	1.008	2382.0
60	0.1995	983.18	0.1304	654.7	21.3	4.183	1.966	465.1	10.81	0.4731	82.89	2.972	0.999	2357.7
70	0.3120	977.75	0.1984	663.5	22.2	4.188	1.987	403.1	11.13	0.4125	56.10	2.546	0.997	2333.1
80	0.4741	971.78	0.2937	670.4	23.1	4.196	2.012	354.5	11.49	0.3648	39.12	2.218	1.000	2308.1
90	0.7018	965.30	0.4239	675.7	24.1	4.205	2.042	314.9	11.87	0.3263	27.99	1.960	1.006	2282.6
100	1.014	958.35	0.5981	679.6	25.1	4.217	2.077	282.4	12.25	0.2947	20.48	1.753	1.014	2256.5
110	1.434	950.95	0.8269	682.2	26.2	4.230	2.121	255.3	12.63	0.2685	15.28	1.583	1.023	2229.7
120	1.987	943.11	1.1220	683.7	27.4	4.246	2.174	232.4	13.00	0.2465	11.59	1.444	1.032	2202.1
130	2.703	934.83	1.4968	684.3	28.7	4.265	2.237	213.0	13.34	0.2279	8.915	1.328	1.040	2173.7
140	3.615	926.13	1.9665	683.9	30.1	4.286	2.311	196.4	13.67	0.2120	6.951	1.231	1.049	2144.2
150	4.761	917.01	2.5478	682.8	31.6	4.310	2.396	182.1	13.97	0.1986	5.485	1.150	1.060	2113.7
160	6.181	907.45	3.2593	680.8	33.2	4.338	2.492	169.8	14.26	0.1871	4.376	1.082	1.072	2081.9
170	7.921	897.45	4.1217	677.9	34.8	4.369	2.599	159.1	14.55	0.1773	3.529	1.025	1.086	2048.7
180	10.03	887.01	5.1583	674.3	36.5	4.406	2.716	149.8	14.83	0.1689	2.875	0.979	1.104	2014.0
190	12.55	876.08	6.3948	669.7	38.3	4.447	2.846	141.6	15.13	0.1616	2.365	0.940	1.125	1977.7
200	15.55	864.67	7.8603	664.3	40.1	4.494	2.990	134.3	15.44	0.1553	1.964	0.908	1.151	1939.7
210	19.07	852.73	9.5875	658.0	42.0	4.548	3.150	127.7	15.77	0.1498	1.645	0.883	1.182	1899.6
220	23.19	840.23	11.614	650.7	44.1	4.611	3.328	121.8	16.13	0.1449	1.389	0.863	1.218	1857.4
230	27.97	827.12	13.984	642.4	46.3	4.683	3.528	116.3	16.51	0.1405	1.181	0.847	1.260	1812.8
240	33.47	813.36	16.748	633.0	48.6	4.767	3.755	111.1	16.91	0.1366	1.010	0.837	1.306	1765.5
250	39.76	798.89	19.965	622.5	51.2	4.865	4.012	106.3	17.33	0.1330	0.8678	0.830	1.356	1715.3
260	46.92	783.62	23.710	610.8	54.1	4.981	4.308	101.7	17.74	0.1297	0.7482	0.829	1.412	1661.8
270	55.03	767.46	28.072	597.7	57.3	5.119	4.655	97.30	18.15	0.1268	0.6465	0.833	1.474	1604.6
280	64.16	750.27	33.163	583.0	60.9	5.286	5.070	93.17	18.55	0.1242	0.5592	0.845	1.544	1543.2
290	74.42	731.91	39.128	566.9	64.9	5.492	5.581	89.26	18.93	0.1220	0.4838	0.865	1.628	1476.8
300	85.88	712.14	46.162	549.1	69.5	5.752	6.223	85.56	19.32	0.1201	0.4185	0.896	1.729	1404.8
310	98.65	690.67	54.529	529.8	75.1	6.088	7.051	82.02	19.74	0.1188	0.3619	0.943	1.853	1325.9
320	112.84	667.08	64.616	509.3	82.3	6.541	8.157	78.53	20.24	0.1177	0.3132	1.009	2.006	1238.6
373.95	220.64	322.00	322.000	1419.0	1419.0			43.16	43.16	0.1340	0.1340			0.0

Source [9.1]

A3: Thermal properties of saturated water and steam (cont.)

Formula for saturation pressure and density of saturated steam
Valid for the total saturation area.

$$p_s = \exp(11.6885 - 3\,746/T - 228\,675/T^2) \cdot \text{bar} \pm 0.8\,\%$$

$$\rho'' = \exp(11.41 - 4\,194/T - 99\,183/T^2) \cdot \text{kg/m}^3 \pm 6\,\%$$

Formula for the range of 0 °C to 320 °C.

$$\sum_{i=0}^{9} C_i \cdot \vartheta_R^{\,i} \qquad \text{with } \vartheta_R = \vartheta / \vartheta_{crit} = \vartheta / 373.95\,°C$$

	ρ'	λ'	λ''	c'_p	c''_p
C_0	999.8	0.561	17.1	4.2196	1.888
C_1	22.92	0.671	−7.55	−1.088	0.341
C_2	−1161	1.283	455.1	10.42	−2.5
C_3	3507	−20.70	−1917	−46.57	68.2
C_4	−9975	79.63	20.5	98.84	−571
C_5	19710	−162.11	26173	55.534	2525
C_6	−25039	185.11	−89184	−709.9	−6074
C_7	18158	−109.56	135700	1454.5	8195
C_8	−6014	22.52	−100790	−1303.6	−5864
C_9	264.9	3.02	29757	451.9	1748
Results in	kg/m³	W/(m K)	10^{-3} W/(m K)	kJ/(kg K)	kJ/(kg K)
Std.-Div. in %	0.017	0.069	0.075	0.056	0.111

	η	η	Pr	Pr	r
C_0	1.791	9.199	13.468	1.0173	2500.9
C_1	−21.595	20.4	−181.25	2.973	−893
C_2	170.01	−208	1510.76	−58.94	224
C_3	−902.200	1499	−8269.4	443.32	−2484
C_4	3201.52	−4926	29914.8	−1729.3	11784
C_5	−7522.95	8326.8	−71208.642	3922.7	−42215
C_6	11487	−7286.3	109707.19	−5359.6	87847
C_7	−10916	3077.8	−104892.38	4367.8	−108004
C_8	5849.41	−655.2	56431.8	−1971.3	72611
C_9	−1347.78	173.29	−13032.7	385.41	−20870
Results in	10^{-3} kg/(m s)	10^{-6} kg/(m s)	-	-	kJ/kg K
Std.-Div. in %	0.017	0.069	0.075	0.056	0.111

Important: Not valid for temperatures over 320 °C !

A4: Thermal properties of water and steam

p bar	ϑ °C	ρ kg/m³	c_P kJ/(kg K)	η 10^{-6} kg/(m s)	ν m²/s	λ 10^{-3} W/(m K)	Pr -
1	0	999.844	4.219	1791.53	1.792	561.08	13.473
	50	988.047	4.180	546.85	0.553	643.61	3.551
	100	0.590	2.074	12.27	20.810	25.08	1.015
	150	0.516	1.986	14.18	27.469	28.86	0.976
	200	0.460	1.976	16.18	35.144	33.28	0.960
	250	0.416	1.989	18.22	43.841	38.17	0.949
	300	0.379	2.012	20.29	53.543	43.42	0.940
	350	0.348	2.040	22.37	64.226	48.97	0.932
	400	0.322	2.070	24.45	75.864	54.76	0.924
2	0	999.894	4.219	1791.28	1.791	561.13	13.468
	50	988.090	4.179	546.87	0.553	643.65	3.551
	100	958.400	4.216	281.77	0.294	679.15	1.749
	150	1.042	2.067	14.13	13.566	29.54	0.989
	200	0.925	2.014	16.15	17.446	33.68	0.965
	250	0.834	2.010	18.20	21.821	38.42	0.952
	300	0.760	2.025	20.28	26.691	43.59	0.942
	350	0.698	2.048	22.36	32.047	49.09	0.933
	400	0.645	2.076	24.45	37.877	54.85	0.925
5	0	1000.047	4.217	1790.53	1.790	561.30	13.454
	50	988.221	4.179	546.92	0.553	643.79	3.550
	100	958.541	4.216	281.85	0.294	679.32	1.749
	150	917.020	4.310	182.47	0.199	682.06	1.153
	200	2.353	2.145	16.05	6.822	34.93	0.986
	250	2.108	2.078	18.14	8.607	39.18	0.962
	300	1.913	2.066	20.24	10.579	44.09	0.948
	350	1.754	2.075	22.34	12.739	49.45	0.938
	400	1.620	2.095	24.44	15.086	55.14	0.929
10	0	1000.301	4.215	1789.28	1.789	561.57	13.430
	50	988.438	4.177	547.01	0.553	644.02	3.548
	100	958.775	4.215	281.99	0.294	679.59	1.749
	150	917.304	4.309	182.59	0.199	682.40	1.153
	200	4.854	2.429	15.89	3.274	37.21	1.037
	250	4.297	2.212	18.05	4.200	40.52	0.985
	300	3.876	2.141	20.19	5.207	44.96	0.961
	350	3.540	2.123	22.31	6.303	50.07	0.946
	400	3.262	2.128	24.42	7.488	55.62	0.935
20	100	959.242	4.212	282.25	0.294	680.14	1.748
	150	917.871	4.305	182.85	0.199	683.07	1.152
	200	865.007	4.491	134.43	0.155	663.72	0.910
	250	8.970	2.560	17.86	1.991	43.49	1.051
	300	7.968	2.320	20.08	2.519	46.82	0.995
	350	7.215	2.230	22.25	3.084	51.37	0.966
	400	6.613	2.200	24.40	3.689	56.62	0.948
	450	6.115	2.196	26.52	4.336	62.32	0.935
	500	5.692	2.207	28.60	5.025	68.34	0.924

Source [9.1]

A4:   Thermal properties of water and steam (cont.)

p bar	ϑ °C	ρ kg/m³	c_P kJ/(kg K)	η 10^{-6} kg/(m s)	ν m²/s	λ 10^{-3} W/(m K)	Pr -
50	200	867.27	4.474	666.329	135.181	0.156	0.908
	250	800.08	4.851	622.501	106.400	0.133	0.829
	300	22.05	3.171	53.848	19.799	0.898	1.166
	350	19.24	2.661	22.127	1.150	55.989	1.052
	400	17.29	2.459	24.369	1.410	60.062	0.998
	450	15.79	2.371	26.550	1.681	65.105	0.967
	500	14.58	2.333	28.681	1.967	70.743	0.946
	550	13.57	2.321	30.766	2.267	76.794	0.930
	600	12.71	2.324	32.810	2.582	83.135	0.917
100	300	715.29	5.682	86.461	0.121	550.675	0.892
	350	44.56	4.012	22.151	0.497	68.088	1.305
	400	37.82	3.096	24.487	0.647	67.881	1.117
	450	33.57	2.747	26.735	0.796	70.987	1.035
	500	30.48	2.583	28.911	0.949	75.607	0.988
	550	28.05	2.501	31.027	1.106	81.106	0.957
	600	26.06	2.460	33.089	1.270	87.139	0.934
	650	24.38	2.442	35.103	1.440	93.478	0.917
	700	22.94	2.438	37.071	1.616	99.978	0.904
200	300	734.71	5.317	90.050	0.123	571.259	0.838
	350	600.65	8.106	69.309	0.115	463.199	1.213
	400	100.51	6.360	26.034	0.259	105.458	1.570
	450	78.62	4.007	27.812	0.354	91.029	1.224
	500	67.60	3.284	29.849	0.442	89.846	1.091
	550	60.35	2.955	31.901	0.529	92.785	1.016
	600	54.99	2.781	33.923	0.617	97.553	0.967
	650	50.78	2.682	35.903	0.707	103.158	0.934
	700	47.32	2.625	37.841	0.800	109.109	0.910
500	300	776.46	4.782	98.477	0.127	618.323	0.762
	350	693.27	5.370	83.236	0.120	541.491	0.825
	400	577.74	6.778	67.983	0.118	451.173	1.021
	450	402.02	9.567	50.477	0.126	315.361	1.531
	500	257.11	7.309	40.499	0.158	202.982	1.458
	550	195.37	5.103	38.690	0.198	163.650	1.206
	600	163.70	4.097	39.121	0.239	151.983	1.055
	650	143.73	3.587	40.249	0.280	149.724	0.964
	700	129.57	3.288	41.648	0.321	150.918	0.907
1000	300	823.18	4.400	109.110	0.133	675.330	0.711
	350	762.34	4.605	95.741	0.126	616.955	0.715
	400	692.92	4.892	84.758	0.122	548.157	0.756
	450	614.19	5.258	74.911	0.122	476.087	0.827
	500	528.20	5.576	66.062	0.125	394.700	0.933
	550	444.48	5.549	59.116	0.133	319.247	1.028
	600	374.22	5.171	54.690	0.146	272.029	1.040
	650	321.08	4.628	52.429	0.163	248.026	0.978
	700	282.00	4.191	51.587	0.183	236.000	0.916

Source [9.1]

A5: Thermal properties of saturated Freon 134a

ϑ	p	ρ'	ρ''	λ'	λ''	c'_p	c''_p	η'	η''	ν'	ν''	Pr'	Pr''	r
°C	bar	kg/m³		10⁻³ W/(m K)		kJ/(kg K)		10⁻⁶ kg/(m s)		10⁻⁶ m²/s		-		kJ/kg
-35	0.662	1403.10	3.521	108.9	8.70	1.264	0.765	405.6	9.507	0.289	2.700	4.71	0.835	222.8
-30	0.844	1388.40	4.426	106.8	9.14	1.273	0.781	381.1	9.719	0.274	2.196	4.54	0.830	219.5
-25	1.064	1373.40	5.506	104.6	9.66	1.283	0.798	358.4	9.946	0.261	1.806	4.39	0.822	216.2
-20	1.327	1358.30	6.785	102.4	10.11	1.293	0.816	337.2	10.160	0.248	1.498	4.26	0.820	213.0
-15	1.639	1342.80	8.287	100.2	10.57	1.304	0.835	317.4	10.380	0.236	1.253	4.13	0.819	209.5
-10	2.006	1327.10	10.041	98.06	11.03	1.316	0.854	298.9	10.590	0.225	1.055	4.01	0.821	206.0
-5	2.433	1311.10	12.077	95.87	11.49	1.328	0.875	281.6	10.810	0.215	0.895	3.90	0.823	202.4
0	2.928	1294.80	14.428	93.67	11.96	1.341	0.897	265.3	11.020	0.205	0.764	3.80	0.827	198.6
5	3.496	1278.10	17.131	91.46	12.43	1.355	0.921	249.9	11.240	0.196	0.656	3.70	0.832	194.7
10	4.146	1261.00	20.226	89.25	12.92	1.370	0.946	235.4	11.460	0.187	0.567	3.61	0.839	190.7
15	4.883	1243.40	23.758	87.02	13.42	1.387	0.972	221.7	11.680	0.178	0.492	3.53	0.846	186.6
20	5.717	1225.30	27.780	84.78	13.93	1.405	1.001	208.7	11.910	0.170	0.429	3.46	0.856	182.2
25	6.653	1206.70	32.350	82.53	14.46	1.425	1.032	196.3	12.140	0.163	0.375	3.39	0.867	177.8
30	7.701	1187.50	37.535	80.27	15.01	1.446	1.065	184.6	12.380	0.155	0.330	3.33	0.879	173.1
35	8.869	1167.50	43.416	77.98	15.58	1.471	1.103	173.4	12.630	0.149	0.291	3.27	0.894	168.2
40	10.165	1146.70	50.085	75.69	16.19	1.498	1.145	162.7	12.890	0.142	0.257	3.22	0.911	163.0
45	11.598	1125.10	57.657	73.37	16.84	1.530	1.192	152.5	13.170	0.136	0.228	3.18	0.932	157.6
50	13.177	1102.30	66.272	71.05	17.54	1.566	1.246	142.7	13.470	0.129	0.203	3.14	0.957	151.8
55	14.913	1078.30	76.104	68.71	18.30	1.609	1.310	133.2	13.790	0.124	0.181	3.12	0.987	145.7
60	16.816	1052.90	87.379	66.36	19.14	1.660	1.387	124.1	14.150	0.118	0.162	3.10	1.030	139.1
65	18.896	1025.60	100.400	64.02	20.09	1.723	1.482	115.2	14.560	0.112	0.145	3.10	1.070	132.0
70	21.167	996.25	115.570	61.69	21.17	1.804	1.605	106.6	15.040	0.107	0.130	3.12	1.140	124.3
75	23.641	964.09	133.490	59.39	22.44	1.911	1.771	98.1	15.600	0.102	0.117	3.16	1.230	115.9
80	26.332	928.24	155.080	57.15	24.00	2.065	2.012	89.7	16.310	0.097	0.105	3.24	1.370	106.4

Source [9.8]

A5:   Thermal properties of saturated Freon 134a (cont.)

Saturation pressure
Validity range: $-35\,°C$ to $+80\,°C$.

$$p_s = \exp(9.94333 - 2137/T - 78124/T^2) \cdot \text{bar} \pm 0.5\,\%$$

Formula for the range of $-35\,°C$ to $+80\,°C$.

$$\sum_{i=0}^{6} C_i \cdot \vartheta_R^i \quad \text{mit } \vartheta_R = \vartheta / \vartheta_{crit} = \vartheta / 101.05\,°C$$

	ρ'	ρ''	λ'	λ''	c_p'	c_p''
C_0	1294.8	14.432	93.661	11.962	1.3412	0.8975
C_1	−333.18	50.335	−44.471	9.46	0.2586	0.4325
C_2	−74.807	71.630	−1.3931	1.37592	0.1849	0.2574
C_3	−2.04	5.0382	0.6391	0.9545	−123.76	2.0974
C_4	−18.472	14.333	−2.9691	−0.9588	−0.0769	−0.1909
C_5	76.9178	−79.744	4.3914	−7.7306	−2.5225	−4.0597
C_6	−159.04	163.240	1.5307	16.077	3.6497	5.9750
Results in	kg/m³	kg/m³	10^{-3} W/(m K)	10^{-3} W/(m K)	kJ/(kg K)	kJ/(kg K)
Std.-Div. %	0.005	0.405	0.013	0.078	0.086	0.184

	η	η	Pr	Pr	r
	10^{-6} kg/(m s)	10^{-6} kg/(m s)	-	-	kJ/kg
C_0	265.27	11.0235	3.7977	0.8271	198.614
C_1	−319.73	4.3362	−2.0036	0.0735	−76.868
C_2	192.61	0.2324	1.4628	0.2551	−26.119
C_3	−123.76	2.0974	−0.2527	0.4615	−14.865
C_4	77.281	0.1718	0.7457	0.1096	0.7774
C_5	−42.575	−5.3073	−2.8296	−3.0793	−11.94
C_6	4.5818	9.1226	3.3467	4.0928	−16.902
Results in	10^{-6} kg/(m s)	10^{-6} kg/(m s)	-	-	kJ/kg
Std.-Div. in %	0.013	0.025	0.063	0.228	0.025

Important: Not valid below $-35\,°C$ and over $+80\,°C$.

A6: Thermal properties of air at 1 bar pressure

ϑ	ρ	λ	c_p	η	ν	Pr	a
°C	kg/m³	10^{-3} W/(m K)	J/(kg K)	10^{-6} kg/(m s)	10^{-6} m²/ s	-	10^{-9} m²/s
−80	1.807	17.74	1009	12.94	7.16	0.7357	9.73
−60	1.636	19.41	1007	14.07	8.60	0.7301	11.78
−40	1.495	21.04	1007	15.16	10.14	0.7258	13.97
−30	1.433	21.84	1007	15.70	10.95	0.7236	15.13
−20	1.377	22.63	1007	16.22	11.78	0.7215	16.33
−10	1.324	23.41	1006	16.74	12.64	0.7196	17.57
0	1.275	24.18	1006	17.24	13.52	0.7179	18.83
10	1.230	24.94	1007	17.74	14.42	0.7163	20.14
20	1.188	25.69	1007	18.24	15.35	0.7148	21.47
30	1.149	26.43	1007	18.72	16.30	0.7134	22.84
40	1.112	27.16	1007	19.20	17.26	0.7122	24.24
60	1.045	28.60	1009	20.14	19.27	0.7100	27.13
80	0.9859	30.01	1010	21.05	21.35	0.7083	30.14
100	0.9329	31.39	1012	21.94	23.51	0.7070	33.26
120	0.8854	32.75	1014	22.80	25.75	0.7060	36.48
140	0.8425	34.08	1016	23.65	28.07	0.7054	39.80
160	0.8036	35.39	1019	24.48	30.46	0.7050	43.21
180	0.7681	36.68	1022	25.29	32.93	0.7049	46.71
200	0.7356	37.95	1026	26.09	35.47	0.7051	50.30
250	0.6653	41.06	1035	28.02	42.11	0.7063	59.62
300	0.6072	44.09	1046	29.86	49.18	0.7083	69.43
350	0.5585	47.05	1057	31.64	56.65	0.7109	79.68
400	0.5170	49.96	1069	33.35	64.51	0.7137	90.38
450	0.4813	52.82	1081	35.01	72.74	0.7166	101.50
500	0.4502	55.64	1093	36.62	81.35	0.7194	113.10
550	0.4228	58.41	1105	38.19	90.31	0.7221	125.10
600	0.3986	61.14	1116	39.17	99.63	0.7247	137.50
650	0.3770	63.83	1126	41.20	109.30	0.7271	150.30
700	0.3576	66.46	1137	42.66	119.30	0.7295	163.50
750	0.3402	69.03	1146	44.08	129.60	0.7318	177.10
800	0.3243	71.54	1155	45.48	140.20	0.7342	191.00
850	0.3099	73.98	1163	46.85	151.20	0.7368	205.20
900	0.2967	76.33	1171	48.19	162.40	0.7395	219.70
1000	0.2734	80.77	1185	50.82	185.90	0.7458	249.20

Source: [9.3]

Formula for the range of −80 °C to 1 000 °C

$$\rho = p / R \cdot T = 348.68 \cdot T^{-1} \cdot \mathrm{K} \cdot \mathrm{kg/m^3} \pm 0.066\ \%$$

$$\sum_{i=0}^{5} C_i \cdot \vartheta_R^i \quad \text{with } \vartheta_R = \vartheta / \vartheta_{crit} = \vartheta / 1000\ \mathrm{K}$$

	λ	c_p	η	ν	Pr	a
C_0	24.18	1006.3	17.23	13.53	0.718	18.84
C_1	76.34	7.4	50.33	89.11	-0.166	128.72
C_2	-48.26	525.6	-34.17	111.36	0.686	168.71
C_3	62.81	-334.5	24.22	-48.80	-0.954	-160.40
C_4	-45.68	-195.2	-4.11	28.60	0.581	155.62
C_5	11.39	175.6	-2.67	-7.91	-0.117	-62.31
Results in	10^{-3} W/(m K)	J/(kg K)	10^{-6} kg/(m s)	10^{-6} m²/s	-	10^{-8} m²/s
Std.-Div. in %	0.013	0.036	0.225	0.062	0.063	0.050

A7:   Thermal properties of solid matter

Metals and alloys

	ϑ	ρ	c_p	λ	a
	°C	kg/m³	J/(kg K)	W/(m K)	$10^{-6} \cdot$ m²/s
Aluminum	20	2700	945	238	93.4
Lead	20	11 340	131	35.3	23.8
Bronze (6 Sn. 9 Zn. 84 Cu. 1 Pb)	20	8 800	377	61.7	18.6
Iron					
Cast iron 3 % C	20	7 870	450	58	14.7
Steel ST 37.8	20	7 830	430	57	16.9
Cr-Ni-Steel 1.4541	20	7 900	470	15	4.1
Cr-Steel X8 Cr7	20	7 700	460	25.1	7.1
Gold (pure)	20	19 290	128	295	119
Copper (pure)	20	8 960	385	394	114

Building materials

Brick wall	20	1 400	840	0.79	0.49
		1 800	840	0.81	0.54
Plaster	20	1 690	800	0.79	0.25
Fir, radial	20	600	2 700	0.14	0.09
Plywood	20	800	2 000	0.15	0.09
Cork tile	30	190	1 880	0.041	0.11
Mineral rock wool	50	200	920	0.064	0.25
Glass wool	0	200	660	0.037	0.28

Stones and glasses

Soil	20	2 040	1 840	0.59	0.16
Fire-brick	100	1 700	840	0.50	0.35
Quartz	20	2 100	780	1.40	0.72
Sandstone	20	2 150	710	1.60	1.00
Marble	20	2 500	810	2.80	1.30
Granite	20	2 750	890	2.90	1.20
Window glass	20	2 480	700	1.16	0.50
Pyrex glass	20	2 240	774	1.06	0.61
Quartz glass	20	2 210	730	1.40	0.87

Plastics

Polyamide	20	1 130	2300	0.280	0.12
Polytetrafluorethyle (Teflon)	20	2 200	1040	0.230	0.10
Rubber, soft	20	1 100	1670	0.160	0.09
Styrofoam	20	15	1250	0.029	0.36
Polyvinyl chloride (PVC)	20	1 380	960	0.150	0.11

Sources [9.1, 9.3]

A8:   Thermal properties of thermal oils

Matter	Producer	Validity range °C	ρ kg/m³	c_p kJ/(kg K)	ν 10^{-6} m²/s	λ W/(m K)
Farolin U	Aral	−10	886	1.80	15.8	0.135
		325	682	3.10	0.60	0.113
Farolin S	Aral	−25	931	1.66	1 396	0.129
		305	710	2.93	0.52	0.113
Farolin T	Aral	−30	914	1.74	91.9	0.132
		300	695	2.84	0.56	0.111
Thermofluid A	AVIA	−25	947	1.70	804	0.133
		250	751	2.68	0.52	0.114
Thermofluid B	AVIA	0	878	1.81	300	0.136
		310	688	2.94	0.59	0.113
Transcal N	BP	0	889	1.95	310	0.135
		320	680	3.04	0.56	0.115
Transcal LT	BP	−20	900	1.80	300	0.136
		260	732	2.77	0.49	0.118
Deacal A 12	Shell & DEA	0	882	1.75	82.6	0.135
		250	720	2.67	0.53	0.117
Deacal 32	Shell & DEA	0	887	1.78	310	0.135
		270	711	2.78	0.68	0.115
Deacal 46	Shell & DEA	0	885	1.80	604	0.133
		280	709	2.81	0.84	0.113
Thermal Oil S	Esso	−10	893	1.80	47.3	0.134
		240	731	2.67	0.52	0.116
Thermal Oil T	Esso	0	877	1.81	285	0.135
		320	670	3.01	0.6	0.112
Essotherm 650	Esso	0	909	1.77	15 803	0.130
		320	702	2.92	1.34	0.108
Caloran 32	Fina	0	883	1.86	300	0.134
		320	648	3.25	0.62	0.111
Mobiltherm 594	Mobil Oil	−44	914	1.64	300	0.135
		250	724	2.70	0.42	0.116
Mobiltherm 603	Mobil Oil	−8	876	1.79	300	0.137
		300	677	2.98	0.52	0.113
Thermia Oil A	Shell	−25	917	1.71	300	0.133
		250	751	2.68	0.52	0.114
Thermia Oil B	Shell	−2	878	1.81	300	0.136
		310	688	2.93	0.59	0.113
Mihatherm WU 10	SRS	−20	914	1.69	341	0.133
		250	752	2.80	0.50	0.113
Mihatherm WU 46	SRS	0	883	1.81	529	0.135
		320	678	2.97	0.60	0.112

Source [9.3]

A9:   Thermal properties of fuels at 1.013 bar

Gasoline

ϑ °C	ρ kg/m³	c_p kJ/(kg K)	λ W/(m K)	η 10^{-6} kg/(m s)	a 10^{-8} m/s	Pr
−50	775	2.051	0.142	0.981	8.89	14.20
−25	755	2.093	0.141	0.686	8.89	10.20
0	735	2.135	0.140	0.510	8.89	7.80
20	720	2.198	0.140	0.402	8.83	6.30
50	690	2.260	0.143	0.294	9.17	4.65
100	650	2.386	0.136	0.196	8.75	3.45

Source [9.3]

Heating Oil S

ϑ °C	ρ kg/m³	c_p kJ/(kg K)	λ W/(m K)	η 10^{-3} kg/(m s)	a 10^{-8} m²/s	Pr -
80	910	2.040	0.1190	67.34	6.41	1155
90	904	2.080	0.1180	44.30	6.28	780
100	898	2.120	0.1170	30.53	6.15	553
110	892	2.160	0.1160	22.30	6.02	415
120	885	2.205	0.1155	16.46	5.92	314
130	879	2.250	0.1150	12.31	5.81	240
140	873	2.280	0.1143	9.34	5.74	186
150	867	2.310	0.1136	7.28	5.67	148
160	861	2.350	0.1129	5.94	5.58	124
170	855	2.390	0.1122	5.13	5.49	109
180	850	2.430	0.1115	4.68	5.40	102

Source [9.1]

A10:   Emissivity of surfaces

1. Metals	Temperature K	ε_n	ε
Aluminum, rolled	443	0.039	0.049
	773	0.050	
-, high-polish	500	0.039	
	850	0.057	
-, oxidized at 872 K	472	0.110	
	872	0.190	
-, strong oxide layer	366	0.200	
	777	0.310	
Aluminum oxide	550	0.630	
	1 100	0.260	
Lead, gray oxidized	297	0.280	
Chrome, high-polish	423	0.058	0.071
	1 089	0.360	
Gold, high-polish	500	0.018	
	900	0.035	
Copper, high-polish	293	0.030	
-, light tarnished	293	0.037	
-, black oxidized	293	0.780	
-, oxidized	403	0.760	
-, scraped	293	0.070	
Inconel, rolled	1 089		0.690
-, sand-blasted	1 089		0.790
Cast iron, polished	473	0.210	
Cast steel, polished	1 044	0.520	
	1 311	0.56	

Oxidized surfaces:

Sheetiron			
-, red rusted	293	0.612	
-, strongly rusted	292	0.685	
-, rolled	294	0.657	
Sheet steel, thick rough oxide layer	297	0.800	
Cast iron, rough surface, strongly oxidized	311 to 522	0.950	
Magnesium, polished	311	0.070	
	811	0.180	
Magnesium oxide	550	0.550	
	1 100	0.200	
Brass, not oxidized	298	0.035	
	373	0.035	
-, oxidized	473	0.610	
	873	0.590	
Nickel, not oxidized	298		0.045
	373		0.060
-, oxidized	473		0.370
	873		0.478

Source [9.3]

A10:   **Emissivity of surfaces** (cont.)

	Temperature K	ε_n	ε
Platinum	422	0.022	
	1 089	0.123	
Quicksilver, not oxidized	298	0.100	
	373	0.120	
Silver, high-polish	311	0.022	
	644	0.031	
Titanium, oxidized	644		0.540
	1 089		0.590
Uranium oxide (U_3O_8)	1 300		0.790
	1 600		0.780
Tungsten	298		0.024
	773		0.071
	1 273		0.150
	1 773		0.230
Galvanized sheet metal			
-, shiny	301	0.228	
-, gray oxidized	297	0.276	

2. Nonmetals

Asbestos, carton	296	0.960	
-, paper	311	0.930	
	644	0.940	
Concrete, rough	273 to 366		
Tar paper	294	0.910	
Plaster	293	0.8 to 0.9	
Glass	293	0.940	
Quartz glass (7 mm thick)	555	0.930	
	1 111	0.470	
Rubber	293	0.920	
Wood, oak, dressed	273 to 366		0.900
-, beech	343	0.940	0.910
Ceramics, fireproof, white Al_2O_3	366		0.900
Carbon, not oxidized	298		0.810
	773		0.790
-, fibre	533		0.950
-, graphite	373		0.760
	773		0.710
Corundum, emery, rough			
	353	0.850	0.840

A10:   **Emissivity of surfaces** (cont.)

3. Coatings, Paints	Temperature K	ε_n	ε
Oil paint, back	366		0.920
-, green	366		0.950
-, red	366		0.970
-, white	366		0.940
Gloss paint, white	373	0.925	
-, matt, black	353	0.970	
Bakelite paint	353	0.935	
Minium paint	373	0.930	
Radiator enamel	373	0.925	
Enamel, white on steel	292	0.897	
Marble, light gray, polished	273 to 366		0.900
Paper	273		0.920
	366		0.940
China, white	295		0.924
Clay, glazed	298		0.900
-, matt	298		0.930
Water	273	0.950	
	373	0.960	
Ice, smooth with water	273	0.966	0.920
-, rough rime	273	0.985	
Brick, red	273 to 366		0.930

Source [9.3]

A11:   Formulary

Energy balance equation

$$V_{KV} \cdot \rho \cdot c_p \frac{d\vartheta}{dt} = \dot{Q}_{12} + \dot{Q}_{Source} + \dot{m} \cdot (h_2 - h_1) \qquad \text{transient}$$

$$\dot{Q}_{12} + \dot{Q}_{Source} + \dot{m} \cdot (h_2 - h_1) = \dot{Q}_{12} + \dot{Q}_{Source} + \dot{m} \cdot (h_2 - h_1) + \dot{m} \cdot c_p \cdot (\vartheta_2 - \vartheta_1) = 0 \quad \text{steady}$$

Rate equations

$$\delta \dot{Q}_{12} = \alpha_2 \cdot (\vartheta_2 - \vartheta_{W2}) \cdot dA_2$$

$$\delta \dot{Q}_{12} = \alpha_W \cdot (\vartheta_{W2} - \vartheta_{W1}) \cdot dA_W$$

$$\delta \dot{Q}_{12} = \alpha_1 \cdot (\vartheta_{W1} - \vartheta_1) \cdot dA_1$$

$$\delta \dot{Q}_{12} = k \cdot (\vartheta_2 - \vartheta_1) \cdot dA_1$$

Conduction in solid bodies

Plane walls

$$\alpha_i = \lambda_i / s_i$$

$$\frac{1}{k} = \frac{1}{\alpha_{f1}} + \sum_{i=1}^{n} \frac{1}{\alpha_{Wi}} + \frac{1}{\alpha_{f2}} = \frac{1}{\alpha_{f1}} + \sum_{i=1}^{n} \frac{s_i}{\lambda_i} + \frac{1}{\alpha_{f2}}$$

$$\frac{\vartheta_{f1} - \vartheta_1}{\vartheta_{f1} - \vartheta_{f2}} = \frac{k}{\alpha_{f1}} \qquad \frac{\vartheta_i - \vartheta_{i+1}}{\vartheta_{f1} - \vartheta_{f2}} = \frac{k}{\alpha_{Wi}} \qquad \frac{\vartheta_2 - \vartheta_{f2}}{\vartheta_{f1} - \vartheta_{f2}} = \frac{k}{\alpha_{f2}}$$

Hollow cylinder

$$\alpha_i = \frac{2 \cdot \lambda_i}{d_{n+1} \cdot \ln(d_{i+1} / d_i)}$$

$$\frac{1}{k} = \frac{d_{n+1}}{d_1} \cdot \frac{1}{\alpha_{f1}} + \sum_{i=1}^{i=n} \frac{d_{n+1}}{2 \cdot \lambda_i} \cdot \ln(d_{i+1} / d_i) + \frac{1}{\alpha_{f2}}$$

$$\frac{\vartheta_{f1} - \vartheta_1}{\vartheta_{f1} - \vartheta_{f2}} = \frac{d_{n+1}}{d_1} \cdot \frac{k}{\alpha_{f1}} \qquad \frac{\vartheta_i - \vartheta_{i+1}}{\vartheta_{f1} - \vartheta_{f2}} = \frac{d_{n+1}}{d_i} \cdot \frac{k}{\alpha_{Wa}} \qquad \frac{\vartheta_{n+1} - \vartheta_{f2}}{\vartheta_{f1} - \vartheta_{f2}} = \frac{k}{\alpha_{f2}}$$

A11: Formulary (cont.)

Hollow sphere

$$\alpha_{Wa} = \frac{2 \cdot \lambda}{d_2 \cdot (d_2 / d_1 - 1)}$$

$$\frac{1}{k} = \frac{d_{n+1}^2}{d_1^2} \cdot \frac{1}{\alpha_{f1}} + \sum_{i=1}^{i=n} \frac{d_{n+1} \cdot (d_{n+1} / d_i - d_{n+1} / d_{i+1})}{2 \cdot \lambda_i} + \frac{1}{\alpha_{f2}}$$

$$\frac{\vartheta_{f1} - \vartheta_1}{\vartheta_{f1} - \vartheta_{f2}} = \frac{d_{n+1}^2}{d_1^2} \cdot \frac{k}{\alpha_{f1}} \qquad \frac{\vartheta_i - \vartheta_{i+1}}{\vartheta_{f1} - \vartheta_{f2}} = \frac{d_{n+1}^2}{d_i^2} \cdot \frac{k}{\alpha_{Wa}} \qquad \frac{\vartheta_{n+1} - \vartheta_{f2}}{\vartheta_{f1} - \vartheta_{f2}} = \frac{k}{\alpha_{f2}}$$

Transient conduction

Dimensionless temperature Θ:
$$\Theta = \frac{\vartheta - \vartheta_\infty}{\vartheta_A - \vartheta_\infty}$$

Biot number:
$$Bi = \alpha \cdot s / \lambda$$

Fourier number:
$$Fo = a \cdot t / s^2$$
See Diagrams 2.11 to 2.13

Contact temperature

$$\vartheta_K = \left(\vartheta_{A1} + \sqrt{\frac{\lambda_2 \cdot \rho_2 \cdot c_{p2}}{\lambda_1 \cdot \rho_1 \cdot c_{p1}}} \cdot \vartheta_{A2} \right) \cdot \left(1 + \sqrt{\frac{\lambda_2 \cdot \rho_2 \cdot c_{p2}}{\lambda_1 \cdot \rho_1 \cdot c_{p1}}} \right)^{-1}$$

Cooling of small body in large bath

$$(\vartheta_1 - \vartheta_{A2}) = (\vartheta_{A1} - \vartheta_{A2}) \cdot e^{-\frac{\alpha \cdot A}{m_1 \cdot c_{p1}} t}$$

Forced convection

Dimensionless characteristic numbers

$$Re_L = \frac{c \cdot L}{\nu} \qquad L = \begin{cases} d_h = 4 \cdot A / U & \text{closed channels} \\ l & \text{plane surfae, } l = \text{length in flow direction} \\ L' = A / U_{proj} & \text{singel body in cross-flow} \end{cases}$$

$$Nu_L = \frac{\alpha \cdot L}{\lambda}$$

$$Pr = \frac{\nu}{a} = \frac{\eta \cdot c_p}{\lambda}$$

A11: Formulary (cont.)

Closed channels

$$Nu_{d_h,turb} = \frac{\xi}{8} \cdot \frac{Re_{d_h} \cdot Pr}{1+12.7 \cdot \sqrt{\xi/8} \cdot (Pr^{2/3}-1)} \cdot \left[1+\left(\frac{d_h}{L}\right)^{2/3}\right] \cdot f_2 \qquad Re_{d_h} = \frac{c \cdot d_h}{\nu} = \frac{\dot{m} \cdot d_h}{A \cdot \eta}$$

$$\xi = \left[1.8 \cdot \log(Re_{d_h}) - 1.5\right]^{-2} \qquad f_2 = \begin{vmatrix} (Pr/Pr_W)^{0.11} \\ (T/T_W)^{0.45} \end{vmatrix}$$

$$Nu_{d_h,lam} = \sqrt[3]{3.66^3 + 0.664^3 \cdot Pr \cdot (Re_{d_h} \cdot d_h/L)^{1.5}} \qquad \gamma = \frac{Re_{d_h} - 2\,300}{7\,700}$$

$$Nu_{d_h} = \begin{vmatrix} Nu_{d_h,lam} & \text{if } Re_{d_h} \le 2\,300 \\ Nu_{d_h,turb} & \text{if } Re_{d_h} \ge 10\,000 \\ (1-\gamma) \cdot Nu_{d_h,lam}(Re_{d_h} = 2\,300) + \gamma \cdot Nu_{d_h,turb}(Re_{d_h} = 10\,000) & \text{otherwise} \end{vmatrix}$$

Plane wall

$$Nu_{L,lam} = 0.644 \cdot \sqrt[3]{Pr} \cdot \sqrt{Re_L}$$

$$Nu_{l,turb} = \frac{0.037 \cdot Re_L^{0.8} \cdot Pr}{1+2.443 \cdot Re_L^{-0.1} \cdot (Pr^{2/3}-1)} \cdot \begin{cases} (Pr/Pr_W)^{0.25} & \text{for liquids} \\ 1 & \text{for gases} \end{cases}$$

$$Nu_l = \sqrt{Nu_{l,lam}^2 + Nu_{l,turb}^2}$$

Single bodies in cross-flow:

$$Nu_{L',0} = \begin{vmatrix} 2 & \text{sphere} \\ 0.3 & \text{cylinder} \end{vmatrix}$$

$$Nu_{L',lam} = 0.664 \cdot \sqrt[3]{Pr} \cdot \sqrt{Re_{L'}} \qquad \text{for } 1 < Re_{L'} < 1000$$

$$Nu_{L',turb} = 0.037 \cdot Re_{L'}^{0.8} \cdot Pr^{0.48} \cdot f_4 \qquad \text{for } 10^5 < Re_{L'} < 10^7$$

$$Nu_{L'} = Nu_{L',0} + \sqrt{Nu_{L',lam}^2 + Nu_{L',turb}^2}$$

$$f_4 = (Pr/Pr_W)^{0.25} \text{ for liquids and } f_4 = (T/T_W)^{0.121} \text{ for gases}$$

A11: Formulary (cont.)

Tube bundles in cross-flow

$a = s_1 / d_a$, $b = s_2 / d_a$ s_1 tube distance perpendicular to flow, s_2 parallel to flow

$$\Psi = \begin{vmatrix} 1 - \dfrac{V_{fest}}{V} = 1 - \dfrac{\pi \cdot d^2 \cdot l}{4 \cdot s_1 \cdot d \cdot l} = 1 - \dfrac{\pi}{4 \cdot b} & \text{for } b \geq 1 \\[3mm] 1 - \dfrac{V_{fest}}{V} = 1 - \dfrac{\pi \cdot d^2 \cdot l}{4 \cdot s_1 \cdot s_2 \cdot l} = 1 - \dfrac{\pi}{4 \cdot a \cdot b} & \text{for } b < 1 \end{vmatrix} \qquad c_\Psi = c_0 / \Psi$$

$$Re_{\Psi, L'} = \frac{c_\Psi \cdot L'}{\nu} \qquad Nu_{\Psi, L'} = Nu_{L'}(Re_{\Psi, L'})$$

$$f_A = \begin{vmatrix} 1 + \dfrac{0.7 \cdot (b/a - 0.3)}{\Psi^{1.5} \cdot (b/a + 0.7)^2} & \text{aligned arrangement} \\[3mm] 1 + \dfrac{2}{3 \cdot b} & \text{staggered arrangement} \end{vmatrix}$$

$$f_n = \begin{vmatrix} 0.74423 + 0.8 \cdot n - 0.006 \cdot n^2 & \text{if } n \leq 5 \\ 0.018 + \exp\left[0.0004 \cdot (n-6) - 1\right] & \text{if } n \geq 6 \end{vmatrix}$$

$$f_j = \begin{vmatrix} 0.6475 + 0.2 \cdot j - 0.0215 \cdot j^2 & \text{if } j < 5 \\ 1 + 1/(j^2 + j) + 3 \cdot (2 \cdot j - 1)/(j^4 - 2 \cdot j^3 + j^2) & \text{if } j > 4 \end{vmatrix}$$

$$Nu_{Bundle} = \alpha \cdot L' / \lambda = Nu_{L'} \cdot f_A \cdot f_n \qquad Nu_{jth\ tube\ row} = \alpha \cdot L' / \lambda = Nu_{L'} \cdot f_A \cdot f_j$$

Finned tubes

$$\dot{Q} = k \cdot A \cdot \Delta\vartheta_m$$

$$\frac{1}{k} = \frac{A}{A_0 + A_{Ri} \cdot \eta_{Ri}} \cdot \frac{1}{\alpha_a} + \frac{d_a}{2 \cdot \lambda_R} \cdot \ln \frac{d_a}{d_i} + \frac{d_a}{d_i} \cdot \frac{1}{\alpha_i}$$

$$\eta_{Ri} = \frac{\tanh X}{X} \qquad\qquad X = \varphi \cdot \frac{d_a}{2} \sqrt{\frac{2 \cdot \alpha_a}{\lambda \cdot s}}$$

Annular fins $\phi = (D / d_a - 1) \cdot \left[1 + 0.35 \cdot \ln(D / d_a)\right]$

Rectangular fins

$$\varphi = (\varphi' - 1) \cdot \left[1 + 0.35 \cdot \ln \varphi\right] \quad \text{with} \quad \varphi' = 1.28 \cdot (b_R / d_a) \cdot \sqrt{l_R / b_R - 0.2}$$

Continous fins

$$\varphi = (\varphi' - 1) \cdot \left[1 + 0.35 \cdot \ln \varphi'\right] \quad \text{with} \quad \varphi' = 1.27 \cdot (b_R / d_a) \cdot \sqrt{l_R / b_R - 0.3}$$

A11: Formulary (cont.)

Straight fins on a plane surfacce $\quad \varphi = 2 \cdot h / d_a$

Tube bundles with annular fins

$$Nu_{d_a} = C \cdot Re_{d_a}^{0.6} \cdot \left[(A_{Ri} + A_0) / A \right]^{-0.15} \cdot Pr^{1/3} \cdot f_4 \cdot f_n$$

$C = 0.2$ aligned, $C = 0.38$ staggered

$$A = \pi \cdot d_a \cdot l \qquad A_0 = \pi \cdot d_a \cdot l \cdot (1 - s / t_R)$$

$$A_{Ri} = 2 \cdot \frac{\pi}{4} \cdot (D^2 - d_a^2) \cdot \frac{l}{t_R}$$

$$\frac{A_{Ri}}{A} = \left[(D / d_a)^2 - 1 \right] \cdot \frac{d_a}{2 \cdot t_R}$$

$$c_e = \begin{vmatrix} c_0 \cdot \left[(1 - \dfrac{1}{a}) - \dfrac{s \cdot (D - d_a)}{s_1 \cdot t_R} \right]^{-1} & \text{if } b \geq 1 \\[3mm] c_0 \cdot \left[\sqrt{1 + (2 \cdot b / a)^2} - \dfrac{2}{a} - \dfrac{2 \cdot s \cdot (D - d_a)}{s_1 \cdot t_R} \right]^{-1} & \text{if } b < 1 \end{vmatrix} \qquad Re = \frac{c_e \cdot d_a}{\nu}$$

Free convection

Vertical walls:

$$Gr = \frac{g \cdot L^3 \cdot \beta \cdot (\vartheta_w - \vartheta_0)}{\nu^2} \quad \text{with} \quad \beta = \frac{1}{T_0} \quad \text{for ideal gases}$$

$$L = A / U_{proj} \qquad Ra = Gr \cdot Pr$$

$$Nu_l = \left\{ 0.825 + 0.387 \cdot (Gr \cdot Pr)^{1/6} \cdot \left(1 + 0.671 \cdot Pr^{-9/16} \right)^{-8/27} \right\}^2$$

Inclined walls

$$Nu_l = \begin{vmatrix} Ra \cdot \cos \alpha & \text{if } Ra \geq Ra_c \\[2mm] 0.56 \cdot (Ra_c \cdot \cos \alpha)^{1/4} + 0.13 \cdot \left[(Ra \cdot \cos \alpha)^{1/3} - Ra_c^{1/3} \right] & \text{if } Ra > Ra_c \end{vmatrix}$$

Horizontal cylinders

$$Nu_{L'} = \left[0.752 + 0.387 \cdot Ra_{L'}^{1/6} \cdot (1 + 0.721 \cdot Pr^{-9/16})^{-8/27} \right]^2$$

A11: Formulary (cont.)

Condensation

Condensation on vertical surfaces and on horizontal tubes

$$L' = \sqrt[3]{\frac{\nu_l^2}{g}} \qquad\qquad \Gamma = \frac{\dot{m}_l}{b} \qquad\qquad Re_l = \frac{\Gamma}{\eta_l}$$

for vertical sufaces, the width of the surface: $b = b$
for vertical tubes, the sum of circumferences: $b = n \cdot \pi \cdot d$
for horizontal tube, the sum of tube lengths: $b = n \cdot l$

Local heat transfer coefficients

$$Nu_{L', \, lam, \, x} = 0.693 \cdot \left(\frac{1 - \rho_g / \rho_l}{Re_l}\right)^{1/3} \cdot f_{well} \qquad Nu_{L', \, turb, \, x} = \frac{0.0283 \cdot Re_l^{7/24} \cdot Pr_l^{1/3}}{1 + 9.66 \cdot Re_l^{-3/8} Pr_l^{-1/6}}$$

$$f_{well} = \begin{cases} 1 & \text{for} & Re_l < 1 \\ Re_l^{0.04} & \text{for} & Re_l \geq 1 \end{cases}$$

$$Nu_{L', \, x} = \frac{\alpha_x \cdot L'}{\lambda_l} = \sqrt{Nu_{L', \, lam, \, x}^{\,2} + Nu_{L', \, turb, \, x}^{\,2}} \cdot (\eta_{ls} / \eta_{lW})^{0.25}$$

Mean heat transfer coefficients

$$Nu_{L', \, lam} = 0.925 \cdot \left(\frac{(1 - \rho_g / \rho_l)}{Re_l}\right)^{1/3} \cdot f_{well} \qquad Nu_{L', \, turb} = \frac{0.020 \cdot Re_l^{7/24} \cdot Pr_l^{1/3}}{1 + 20.52 \cdot Re_l^{-3/8} Pr_l^{-1/6}}$$

$$Nu_{L'} = \frac{\alpha \cdot L'}{\lambda_l} = \sqrt[1.2]{Nu_{L', \, lam}^{\,1.2} + Nu_{L', \, turb}^{\,1.2}} \cdot (\eta_{ls} / \eta_{lW})^{0.25}$$

Condensation in tubes with steam flow

Local heat transfer coefficients with downward steam flow

$$Nu_{L', \, x}^* = (1 + \tau_{ZP}^*)^{1/3} \cdot \sqrt{(C_{lam} \cdot Nu_{L', \, lam, \, x})^2 + (C_{turb} \cdot Nu_{L', \, turb, \, x})^2}$$

$$\tau_g^* = \frac{\tau_g}{g \cdot \rho_l \cdot \delta^+} \qquad \tau_g = \frac{\zeta_g \cdot \rho_g \cdot \bar{c}_g^2}{8} \qquad \zeta_g = 0.184 \cdot Re_g^{-0.2} \qquad Re_g = \frac{\bar{c}_g \cdot d_i}{\nu_g}$$

$$\tau_{ZP}^* = \tau_g^* \cdot [1 + 550 \cdot F \cdot (\tau_{ZP}^*)^a] \qquad a = \begin{cases} 0.30 & \text{for} & \tau_g^* \leq 1 \\ 0.85 & \text{for} & \tau_g^* > 1 \end{cases}$$

A11: Formulary (cont.)

$$F = \frac{\max\left[(2 \cdot Re_l)^{0.5}; 0.132 \cdot Re_l^{0.9}\right]}{Re_g^{0.9}} \cdot \frac{\eta_l}{\eta_g} \cdot \sqrt{\frac{\rho_g}{\rho_l}}$$

$$C_{lam} = 1 + (Pr_l^{0.56} - 1) \cdot \tanh(\tau_{ZP}^*) \qquad C_{turb} = 1 + (Pr_l^{0.08} - 1) \cdot \tanh(\tau_{ZP}^*)$$

$$\frac{\delta^+}{d} = \frac{6.59 \cdot F}{\sqrt{1 + 1\,400 \cdot F}}$$

Local heat transfer coefficients with upward steam flow

$$\tau_{ZP}^* = \tau_g^* \cdot [1 + 1\,400 \cdot (\tau_{ZP}^*)^a]$$

$$We = \frac{\tau_{ZP} \cdot \delta^+}{\sigma_l} = \frac{\tau_{ZP}^* \cdot \rho_l \cdot g \cdot (\delta^+)^2}{\sigma_l} \quad \text{must be smaller than 0.01}$$

Local heat transfer coefficients in horizontal tubes

$$Nu_{L', x}^* = \tau_g^{*1/3} \cdot \sqrt{(C_{lam} \cdot Nu_{L', lam, x})^2 + (C_{turb} \cdot Nu_{L', turb, x})^2}$$

$$\varepsilon = 1 - \frac{1}{1 + \dfrac{1}{8.48 \cdot F}} \qquad \delta = 0.25 \cdot (1 - \varepsilon) \cdot d_i \qquad c_g = \frac{4 \cdot \dot{m} \cdot x}{\rho_g \cdot \pi \cdot (d_i - 2 \cdot \delta)^2}$$

$$\tau_g = \frac{0.184 \cdot Re_g^{-0.2}}{8} \cdot c_g^2 \cdot \rho_g \qquad \tau_g^* = \frac{\tau_g}{g \cdot \rho_l \cdot \delta} \cdot (1 + 850 \cdot F)$$

Boiling

Nucleate boiling

$$d_A = 0.0149 \cdot \beta^0 \cdot \sqrt{\frac{2 \cdot \sigma}{g \cdot (\rho_l - \rho_g)}} \qquad \alpha_B = \frac{\dot{q}}{\vartheta_W - \vartheta_s} = \frac{\dot{q}}{\Delta\vartheta} \qquad Nu_{d_A} = \frac{\alpha_B \cdot d_A}{\lambda_l}$$

Water: $b^0 = 45°$
Freon: $b^0 = 35°$
Benzene: $b^0 = 40°$

$$\alpha_B = \alpha_0 \cdot f(p^*) \cdot \left(\frac{\lambda_l \cdot \rho_l \cdot c_{pl}}{\lambda_{l0} \cdot \rho_{l0} \cdot c_{pl0}}\right)^{0.25} \cdot \left(\frac{R_a}{R_{a0}}\right)^{0.133} \cdot \left(\frac{\dot{q}}{\dot{q}_0}\right)^{0.9 - 0.3 \cdot p^*}$$

$$p^* = p / p_{krit} \qquad R_{a0} = 0.4 \ \mu m \qquad \dot{q}_0 = 20\,000 \ W/m^2 \qquad p_0^* = 0.1$$

A11: **Formulary** (cont.)

$$f(p^*) = \begin{cases} 1.73 \cdot p^{*0.27} + \left(6.1 + \dfrac{0.68}{1-p^*}\right) \cdot p^{*2} & \text{for water} \\[3mm] f(p^*) = 1.2 \cdot p^{*0.27} + \left(2.5 + \dfrac{1}{1-p^*}\right) \cdot p^* & \text{for other pure liquids} \end{cases}$$

$$Nu_{d_A 0} = 0.1 \cdot \left(\frac{\dot{q}_0 \cdot d_A}{\lambda_l \cdot T_s}\right)^{0.674} \cdot \left(\frac{\rho_g}{\rho_l}\right)^{0.156} \cdot \left(\frac{r \cdot d_A^2}{a_l^2}\right)^{0.371} \cdot \left(\frac{a_l^2 \cdot \rho_l}{\sigma \cdot d_A}\right)^{0.35} \cdot Pr_l^{-0.16}$$

$$\alpha_0 = \frac{f(0.1)}{f(0.03)} \cdot Nu_{d_A 0} \cdot \frac{\lambda_l}{d_A} = \frac{1}{f(0.03)} \cdot Nu_{d_A 0} \cdot \frac{\lambda_l}{d_A}$$

Boiling of flowing fluids

$$Re_l = \frac{c_{0l} \cdot d_h}{v_l} = \frac{\dot{m} \cdot d_h}{A \cdot \eta_l} \qquad Re_g = \frac{c_{0g} \cdot d_h}{v_l} = \frac{\dot{m} \cdot d_h}{A \cdot \eta_g} \qquad R = \rho_l / \rho_g$$

$$\bar{\alpha} = \frac{1}{x_2 - x_1} \cdot \int_{x_1}^{x_2} \alpha(x)\,dx$$

Vertical tubes

$$\frac{\alpha_x}{\alpha_{l0}} = \left\{ \begin{array}{l} (1-x)^{0.01} \cdot \left[(1-x)^{1.5} + 1{,}9 \cdot x^{0.6} \cdot R^{0.35}\right]^{-2.2} + \\[3mm] + x^{0.01} \cdot \left[\dfrac{\alpha_{g0}}{\alpha_{l0}}\left(1 + 8 \cdot (1-x)^{0.7} \cdot R^{0.67}\right)\right]^{-2} \end{array} \right\}^{-0.5}$$

Horizontal tubes

$$\frac{\alpha_x}{\alpha_{l0}} = \left\{ \begin{array}{l} (1-x)^{0.01} \cdot \left[(1-x)^{1.5} + 1.2 \cdot x^{0.4} \cdot R^{0.37}\right]^{-2.2} + \\[3mm] + x^{0.01} \cdot \left[\dfrac{\alpha_{g0}}{\alpha_{l0}}\left(1 + 8 \cdot (1-x)^{0.7} \cdot R^{0.67}\right)\right]^{-2} \end{array} \right\}^{-0.5}$$

A11: **Formulary** (cont.)

Thermal radiation

$$\dot{Q}_{12} = C_{12} \cdot A \cdot \left[\left(\frac{T_1}{100} \right)^4 - \left(\frac{T_2}{100} \right)^4 \right] \qquad C_s = 5.67 \ \text{W} \cdot \text{m}^{-2} \cdot \text{K}^{-4}$$

Two parallel plates of equal size

$$C_{12} = \frac{C_s}{1/\varepsilon_1 + 1/\varepsilon_2 - 1}$$

Multiple parallel plates of equal size

$$C_{12} = \frac{C_s}{1/\varepsilon_1 + 1/\varepsilon_2 - 1 + \sum_{i=1}^{n} (1/\varepsilon_{i1} + 1/\varepsilon_{i2} - 1)}$$

Enclosed bodies

$$C_{12} = \frac{C_s}{\dfrac{1}{\varepsilon_1} + \dfrac{A_1}{A_2} \cdot \left(\dfrac{1}{\varepsilon_2} - 1 \right)}$$

Bibliography

[1.1] Böckh P von, Cizmar J, Schlachter W (1999) Grundlagen der technischen Thermodynamik. Aarau. Bildung Sauerländer, Aarau; Fortis-Verl. FH, Mainz, Köln

[1.2] Schlünder E-U und Martin H (1995) Einführung in die Wärmeübertragung, 8. neu bearbeitete Aufl, Vieweg Verlag, Braunschweig, Wiesbaden

[1.3] Nußelt W (1915) Das Grundgesetz des Wärmeüberganges. Gesundh. Ing. Bd 38, pp. 477-482 and pp. 490-496

[1.4] Fourier J (1822) Theorie analytique de la chaleur. Didot, Paris, Nachdruck: Paris: Éditions J. Gabay 1988. See also: Grattan-Guinness, Ravetz I u. JR: Joseph Fourier, pp. 1768–1830. Cambridge, MA (USA): MIT Pr 1972

[1.5] Wagner W (1991) Wärmeübertragung, 3. Aufl, Vogel, (Kamprath-Reihe), Würzburg

[1.6] Böckh P von (2001) Fluidmechanik. 2. Aufl, Springer, Berlin, Heidelberg, New York

[1.7] VDI-Wärmeatlas (2002), 9. Aufl, Springer, Berlin, Heidelberg, New York

[2.1] VDI-Wärmeatlas (2002), 9. Aufl, Springer, Berlin, Heidelberg, New York

[2.2] Grigull U und Sandner H. (1979) Wärmeleitung. Springer, Berlin, Heidelberg, New York

[2.3] Martin H und Saberian M (1992) Verbesserte asymptotische Näherungsgleichungen zur Lösung instationärer Wärmeleitungsprobleme auf einfachste Art. Vortrag GVC Fachausschuss Wärme- u. Stoffübertragung Mai 1992, Baden-Baden

[3.1] Wagner W (1991) Wärmeübertragung, 3. Aufl, Vogel, (Kamprath-Reihe), Würzburg

[3.2] Böckh P von (2001) Fluidmechanik. 2. Aufl, Springer, Berlin, Heidelberg, New York

[3.3] Gnielinski V (1995) Forsch im Ing-Wes 61, 9, pp. 240/248

[3.4] VDI-Wärmeatlas (2002), 9. Aufl, Springer, Berlin, Heidelberg, New York

[3.5] Petukhov BS, Roizen LI (1964) High Temperature 2, pp. 65/68

[3.6] Gnielinski V und Gaddis ES (1978) Berechnung des mittleren Wärmeübergangskoeffizienten im Außenraum von Rohrbündel-Wärmetauschern mit Umlenkblechen. Verfahrenstechnik 12, 4, pp. 211/217

[3.7] Paikert P und Schmidt KG (1990) Arbeitsbericht. Fachgeb. Verfahrenstechnik, Universität GH Duisburg, Sept. 1990

[3.8] Briggs DE and Young EH (1963) Eng. Prog Sym Ser Vol 59 No. 41, pp. 1/9

[4.1] VDI-Wärmeatlas (2002), 9. Aufl, Springer, Berlin, Heidelberg, New York

[4.2] Churchill SW (1977) A comprehensive correlating equation for laminar, assisting, forced and free convection. AIChE Journal Vol 23, No. 1, pp. 10/16

[5.1] Nußelt W (1916) Die Oberflächenkondensation des Wasserdampfes. VDI-Zeitschriften, 60, 27

[5.2] Müller J (1992) Wärmeübergang bei der Filmkondensation und seine Einordnung in Wärme- und Stoffübergangsvorgänge bei Filmströmungen. Fortsch. Ber. VDI, Reihe 3, Nr 270

[5.3] Numrich R (1990) Influence of Gas Flow on Heat Transfer in Film Condensation. Chem Eng Technol 13, pp. 136/143

[5.4] Blanghetti F (1979) Lokale Wärmeübergangszahlen bei der Kondensation mit überlagerter Konvektion in vertikalen Rohren. Dissertation, Universität Karlsruhe

[5.5] Rohsenow WM, Hartnett JP, Ganic EN (1985) Handbook of Heat Transfer Fundamentals. 2nd Edition, McGraw Hill, New York

[5.6] VDI Heat Atlas, (2010) 2nd Edition, VDI-Gesellschaft, Düsseldorf

[5.7] Chawla JM and Wiskot G (1992) Wärmeübertragung, Berechnungen mit dem PC, VDI-Verlag GmbH, Düsseldorf

[6.1] Stephan K and Abdelsalam M (1963) Heat-transfer Correlations for Natural Convection Boiling. Int J Heat Mass Transfer, Vol 6, pp. 73/87

[6.2] Stephan K und Preußer P (1979) Wärmeübergang und maximale Stromdichte beim Behältersieden binärer und ternärer Flüssigkeitsgemische. Chem-Ing Techn MS 649/79, Synops Chem-Ing Techn 51, p. 37

[6.3] Steiner D (1982) Wärmeübergang bei Strömungssieden von Kältemitteln und kryogenen Flüssigkeiten in waagerechten und senkrechten Rohren. DKV-Tagungsbericht, Essen, 9. Jahrgang, pp. 241/260

[6.4] Chawla JM (1967) Wärmeübergang und Druckabfall in waagerechten Rohren bei der Strömung von verdampfenden Kältemitteln. VDI-Forsch-Heft 523. Düsseldorf: VDI-Verlag

[7.1] Blevin WR and Brown WJ (1971) A Precise Measurement of the Stefan-Boltzmann Constant. Metrologica 7, pp. 15/29

[7.2] VDI-Wärmeatlas (2002), 9. Aufl, Springer, Berlin, Heidelberg, New York

[7.3] Siegel R, Howell JR, Lohrengel J (1988) Wärmeübertragung durch Strahlung. Teil 1, Springer, Berlin, Heidelberg, New York

[7.4] Wagner W (1991) Wärmeübertragung, 3. Aufl Vogel, (Kamprath-Reihe), Würzburg

[7.5] Gubareff, Jansen, Torborg (1960) Thermal Radiation Properties. Honeywell Research Center, Minneapolis

[7.6] Tables of Emissivity of Surfaces. (1962) Int J Heat & Mass Transfer 5, pp. 67/76

[8.1] Martin H (1988) Wärmeübertrager. Thieme, Stuttgart, New York

[8.2] Roetzel W und Spang B (1990) Verbessertes Diagramm zur Berechnung von Wärmeübertragern. Wärme- und Stoffübertragung 25, pp. 259/264

[8.3] Spang B und Roetzel W (1995) Neue Näherungsgleichung zur einheitlichen Berechnung von Wärmeübertragern. Heat Mass Transf 30, pp. 417/422

[8.4] Roetzel W, Heggs PJ, Butterworth D (Eds.) (1992) Design and Operation of Heat Exchangers. Proc. Eurotherm Seminar No. 18, Hamburg, Feb 27-March 1, 1991, Springer, Berlin, Heidelberg, New York, pp. 19/29

[8.5] VDI-Wärmeatlas (2002), 9. Aufl, Springer, Berlin, Heidelberg, New York

[8.6] TEMA Standards Of The Tubular Heat Exhangers Manufacturers Assotiation, 9th Edition, Section 6

[9.1] Wagner W (1998) Software zur Industrieformulation IAPWS-IF97 zur Berechnung der Zustandsgrößen von Wasser und Wasserdampf, Ruhr Universität Bochum

[9.2] ICI KLEA (1988) Programm Clea Calc Windows, Version 4.0

[9.3] VDI-Wärmeatlas (2002), 9. Aufl, Springer, Berlin, Heidelberg, New York

[9.4] Software zur Industrie-Formulation IAPWS-IF97 von Prof. Dr.-Ing. W. Wagner, Ruhr-Universität Bochum, Lehrstuhl für Thermodynamik, Bochum

[9.5] VDI-RICHTLINIE. (2003) VDI 4670: Thermodynamische Stoffwerte von feuchter Luft und Verbrennungsgasen

[9.6] Wagner W und Kretzschmar H-J (1998) Zustandsgrößen von Wasser und Wasserdampf, IAPWS-IF97, Springer-Verlag, Berlin

[9.7] Baehr H D und Tiller-Roth R(1995) Thermodynamische Eigenschaften umweltverträglicher Kältemittel, Springer-Verlag, Berlin

Index